Coping with Water Deficiency

ENVIRONMENT & POLICY

VOLUME 48

The titles published in this series are listed at the end of this volume

Phoebe Koundouri
Editor

Coping with Water Deficiency

From Research to Policymaking

With Examples from Southern Europe, the Mediterranean and Developing Countries

Springer

Dr Phoebe Koundouri (BA, MPhil, MSc, PhD)
Assistant Professor in Economics
DIEES, Athens University of Economics and Business
76, Patission Street
Athens 104 34
Greece
E-mail: pkoundouri@aueb.gr
Web Page: http://www.phoebekoundouri.com

Cover Illustration: Characteristic of coastal arid in the south mediterannean region, photo by Phoebe Koundouri

ISBN: 978-1-4020-6614-6 e-ISBN: 978-1-4020-6615-3

Library of Congress Control Number: 2007938215

Printed on acid-free paper.

9 8 7 6 5 4 3 2 1

springer.com

As always, to Nikitas, and for the first time to our little Chrisilou, who came to this world when I was editing this book.

Also, to my parents; they know why.

Acknowledgements

My debt is to the European Commission DG Research for financially supporting the ARID Cluster of projects and to the responsible scientific officer, Dr. Panagiotis Balabanis, for trusting me with the coordination of the 80 researchers and the three interdisciplinary European Research projects (Aquadapt, MEDIS, WaterStrategy Man) that participated in this cluster.

I am also grateful to all the contributing authors of this book, for sharing their research results and worldwide experiences in the field of water economics and management. Moreover, I owe a special intellectual debt to all my colleagues from the Athens University of Economics and Business, and also to my co-authors from the University of Cambridge, University of Toulouse, University College London, University of Reading, University of California, San Diego, OECD and the World Bank. I thank them for providing me with a stimulating and challenging academic environment.

I also wish to thank my publisher, Springer, whose staff were enthusiastic and helpful about this book since proposal stage. I owe a special thanks to Tamara Welschot and Judith Terpos.

Contents

Contributors

Bernard Barraqué
Chair of the French UNESCO/IHP Hydrological Committee LATTS-ENPC, Cité Descartes, France. E-mail: barraque@mail.enpc.fr

Ekin Birol
Research Fellow, Department of Land Economy, University of Cambridge, Hills Road, Cambridge, CB2 2PH, UK. E-mail: eb337@cam.ac.uk

Ariel Dinar
Principal Economist, World Bank, Washington DC, USA.
E-mail: ADinar@worldbank.org

Antonia Donta
Senior Researcher, Centre for Environmental Research, Westfälische Wilhelms-Universität Münster, Röntgenstraße 17, D-48149, Münster, Germany. E-mail: Dr.Donta@gmx.de

Carlo Giupponi
Professor, Università Statale di Milano, Dipartimento di Produzione Vegetale, Via Celoria, 2 I-20133 Milano Italia; and Principal Researcher, Fondazione Eni Enrico Mattei, Campo S. Maria Formosa, Castello 5252, 30122 Venice, Italy.
E-mail: carlo.giupponi@feem.it

Ben Groom
Lecturer, SOAS, University of London, London, UK.
E-mail: bg3@soas.ac.uk

Paul Jeffrey
Principal Research Fellow, Centre for Water Science, Cranfield University. UK. E-mail: p.j.jeffrey@cranfield.ac.uk

Christos Karavitis
Lecturer, Agricultural University of Athens, Athens, Greece.
E-mail: ckaravitis@aua.gr

Pipina Katsiardi
Researcher, National Technical University of Athens, Athens, Greece.
E-mail: katsiard@chemeng.ntua.gr

Phoebe Koundouri
Assistant Professor, DIEES, Athens University of Economics and Business, 76 Patission str., Athens 10434, Greece. E-mail: pkoundouri@aueb.gr

Yiannis Kountouris
Researcher and PhD candidate, DIEES, Athens University of Economics and Business, 76 Patission str., Athens 10434, Greece.
E-mail: ykountouris@aueb.gr

Kyriaki Remoundou
Researcher and PhD candidate, DIEES, Athens University of Economics and Business, 76 Patission str., Athens 10434, Greece.
E-mail: Kyriakiremoundou@yahoo.gr

Xiaoyimg Liu
Department of Economics, Normal University, Beijing, China.
E-mail: x.liu@bnu.edu.cn

James McGlade
Assistant Professor, Dept. Humanitats de la UPF. Autonomous University of Barcelona, Spain. Email: j.mcglade@aub.es

Brian S. McIntosh
Centre for Water Science, Cranfield University. UK.
E-mail: b.mcintosh@cranfield.ac.uk

Lange Manfred
Professor, Director, Institute for Geophysics and Centre for Environmental Research, Westfälische Wilhelms-Universität Münster, Corrensstraße 24, D-48149, Münster, Germany. E-mail: langema@uni-muenster.de

The MEDIS consortium consists of the following Institutions: Centre for Environmental Research, University of Muenster, Germany; Institute of Geoinformatics, University of Muenster, Germany; Institute of Geophysics University of Muenster, Germany; Centre of Ecology and Hydrology, National Environment Research Council, UK; Department of Political and Social Sciences, University of Cyprus, Cyprus; Institute of Electronic Structure and Laser, Foundation for Research and Technology, Hellas, Greece; NAGREF, Subtropical Plants and Olive Tree Institute, Greece; Regional Governor of Crete, Water Resources Management Department, Greece; Dipartimento di Construzioni e Tecnologie Avanzate, University of Messina, Italy; Système Physique de l'Environment – URA CNRS 2053, Université de Corse, France; Dep. De Ingenieria del Terreno, Universitat Polytecnica de Catalonya; Balearic Island University Spain; for further details, please see: http://www.uni-muenster.de/Umweltforschung/medis/index.html

Susanne Neubert
Researcher, German Development Institute, Bonn, Germany.
E-mail: sihem.benabdallah@inrst.rnrt.tn

Maria R. Saleth
Researcher, International Water Management Institute, Colombo,
Sri Lanka. E-mail: r.salethr@cgiar.org

Alessandra Sgobbi
Researcher, Fondazione Eni Enrico Mattei, Campo S. Maria Formosa,
Castello 5252, 30122 Venice, Italy. E-mail: Alessandra.sgobbi@feem.it

Timothy Swanson
Chair of Economics and Law, Department of Economics and School of Public
Policy, University of London, London, UK. E-mail: timothy.swanson@ucl.ac.uk

Chapter 1
Introduction

Phoebe Koundouri, Yiannis Kountouris, and Kyriaki Remoundou

Aridity and water stress are global problems with far-reaching economic and social implications. Furthermore there is evidence that the world is enduring a serious water crisis. Its causes can be traced in the unsustainable management of water resources, and water scarcity from effects of the natural environment and increasing demand patterns. Given the numerous and increasing pressures on water resources, due to climatic conditions, over-exploitation of existing surface and ground waters, insufficient recharge due to diminishing precipitation, excessive water use by agricultural activities or tourism and conflicting interests between various users it is vital that effective legislation clearly address the problems and help secure these resources for future generations.

These issues are even more acute in many arid and semi-arid regions in Southern Europe, the Mediterranean, as well as in developing countries, which are characterized by high spatial and temporal imbalances of water demand and supply, seasonal water uses, inadequate water resources and poor institutional water management. There, the need for appropriate strategies and guidelines for water management are necessary for the formulation and implementation of sustainable water resource management.

The multitude and variety of problems faced by water resources worldwide, emerging from sources as diverse as environmental and socioeconomic conditions, stress the need for the implementation of strategies for integrated water resources management (IWRM) in water-deficient regions. Integrated Water Resources Management could achieve sustainable social and economic development combined with the protection of natural ecosystems and the aquatic environment. Thus management of this most precious resource would be accomplished by an interdisciplinary methodology that takes into account environmental and economic parameters to ensure that the resulting guidelines may be applicable to the entire range of conditions found in water-deficient regions.

In this respect, the Water Framework Directive (WFD) developed jointly by the Member States and the European Commission and agreed in May 2001, adopts a holistic approach towards sustainable water resource management and sets clear objectives that a "good status" must be achieved for all European waters by 2015. In essence, the Framework Directive aims to prevent pollution at source and sets

P. Koundouri (ed.), *Coping with Water Deficiency*, 1–7.

out control mechanisms to ensure that all pollution sources are managed in a sustainable way. The Directive also aims at providing sustainable solutions to water scarcity problems that derive from unmanaged demand. Finally, the Directive also requires cooperation across countries and encourages citizens and stakeholders such as NGOs and public authorities at all levels of government to get more involved to ensure that its ambitious objectives will be met within the given deadlines.

The first aim of the book is to report the culmination of the results of the ARID cluster of projects which examine water scarcity and demand in arid and semi-arid regions, as well as participatory and adaptive approaches for appropriate management strategies in the Mediterranean and Southern European Countries. In this sense this is the second book in a series of two, the first entitled: *Water Management in Arid and Semi-Arid Regions: Interdisciplinary Perspectives*. In order to stress the similarities and differences in water management practices and enhance the understanding of their problems around the world, this volume also reports recent examples of methods for the identification of water management issues and the proposed mitigation measures in developing countries, which were produced outside the ARID cluster. A wide spectrum of topics important to water resources management is covered, including research tools for the characterization of water stress, as well as policy proposals for its alleviation, tools and methods for decision making, water policies and pricing schemes, stakeholder participation and social issues. The project's outcomes, lessons learned and conclusions reached are incorporated and discussed in an attempt to formulate water management policies appropriate for arid and semi-arid regions in the context of implementing the Water Framework Directive. Emphasis is given on the potential transfer of research results into concrete policy recommendations towards the development of integrated water resources management strategies acceptable to the communities in Mediterranean countries.

The ARID Cluster of projects aimed at consolidating the work of three EU-funded projects with a view at ensuring that through collaboration, information sharing and dissemination, a consistent set of recommendations, user-friendly tools and methodologies for water management in arid and semi-arid areas are developed. In addition, the Cluster attempted to outline the unique characteristics of aridity and related concepts and stress the demand for comprehensive water resources planning and management as these are proposed in the European Water Framework Directive. The three projects that formed the Cluster were:

- WaterStrategyMan which aimed to develop an integrated water resources management framework to meet the EU requirements concerning preservation and enhancement of the water quality in water-deficient regions
- AQUADAPT which aimed to generate knowledge supporting the strategic planning and management of water resources in semi-arid environments at catchment level under changing supply/demand patterns
- MEDIS which focused on the specification of recommendations for sustainable use of water on islands of the Mediterranean (Corsica, Crete, Cyprus, Mallorca, Sicily) where conflicting demand for water is combined with a wide range of hydrological, social and economic conditions

The volume is organized as follows: Part I includes Chapters 2–6, which examine case studies from Southern Europe and the Mediterranean and provide the results and conclusions from research done under the ARID Cluster, together with Chapter 7, which presents research results carried out with partial financial support from the European Commission projects MULINO, TRANSCAT, NOSTRUM-DSS, and of the Italian Ministry for University and Research. Part II, is composed of Chapters 8 and 9, and provides insights and research results from the developing world. The research presented in Part II of the book was funded by the World Bank Research Committee, the International Water Management Institute and the China Council for International Cooperation for Environment and Development. The chapters of the book present broad and general concepts related to water deficiency management and shed light on ways to choose among potential management options.

In Chapter 2, Antonia Donta, Manfred Lange and the MEDIS consortium argue for integrated water resources management stressing the importance of merging the approaches of natural scientists and engineers with the consideration of social and economic aspects. In line with this, authors suggest the examination of specific natural and socioeconomic indicators as analytical tool in order to derive the most appropriate recommendations for water management on the Mediterranean islands and the wider Mediterranean area by consulting local stakeholders as a key of the approach. Such indicators facilitate the characterization of an integrated picture of water management conditions and the existing situation, and render the comparison between different situations and catchments possible. By using indicators and following the Driving forces-Pressures-State-Impacts-Responses approach, both environmental, i.e., pressures exerted by natural phenomena, and the socio-economic stressors, i.e., conditioned by anthropogenic activities, can be considered. Stakeholder involvement was secured via workshops and discussion with the stakeholders. Responses were not given in a technical way but were rather derived through stakeholder consultation and the process was expanded beyond political and scientific limitations. This chapter shows how this approach was carried out in the MEDIS-project conducted on the islands of Corsica, Crete, Cyprus, Majorca and Sicily. For each of the MEDIS case studies natural environment, agricultural, water quality and socioeconomic indicators including aridity index, water availability and exploitability, gross, person and sector water consumption are presented. A comparison of the islands' water management situations is carried out based on these indicators and proposals are given for a more sustainable and sound water management. These proposals contribute directly to the implementation of the Water Framework Directive (WFD) on the Mediterranean basin.

The precise characterization of the circumstances relating to water supply and demand in arid and semi-arid Mediterranean and Southern European countries is crucial for appreciating the nature of the problems pertaining to water availability and use. This in turn could facilitate the development of policies and actions tailored to the needs of each specific site for the implementation of the Water Framework Directive. In this spirit, in Chapter 3, Bernard Barraqué, Christos Karavitis and Pipina Katsiaridi descriptively present the range of existing circumstances in a number of case studies, carried out by the WaterStrategyMan project. Information relating to climatic conditions, water availability, water quantity and quality, water use and demand, pricing system, water resources management and water policy development priorities is presented for 15 regions selected from six countries. These include Attica, Thessaly and the Cyclades Islands in Greece, Belice basin and Emilia-Romagna in Italy, Doñana and the Canary Islands in Spain, Algarve, Sado and Guadiana in Portugal, Akrotiri, Germasogeia and Kokkinochoria in Cyprus and Tel Aviv and Arava in Israel. For each of the case studies region-specific summary matrixes of descriptive indices are constructed related to the prevailing natural conditions and infrastructure, economic and social system, and decision-making processes.

The conservation and future sustainability of vulnerable fluvio-coastal environments, along with the need for viable planning criteria and policy instruments for their long-term management, are some of the central issues at the heart of the contemporary environmental discourse. In addressing a number of water management related issues, Chapter 4 by James McGlade, Brian S. McIntosh and Paul Jeffrey focus on resilience as a manifestation of sustainability and the notion of "landscape sensitivity", assessing its usefulness as a theoretical construct that might contribute to a better understanding of watershed dynamics, in climatically marginal environments. Perhaps the most significant barrier to interventionist strategies for sustainable water resource management stems from the lack of holistic thinking at governmental and managerial levels resulting in further increase in pollution, soil erosion, pressure on water consumption and general degradation of the environment including its cultural and natural heritage. Authors take a critical look at the theoretical basis within which current research on socio-natural systems is undertaken and express their opposition towards the utilitarian philosophy which, to their view, underpins the recent environmental valuation approaches. Further this chapter stresses the importance of considering moral and ethical issues in the environmental debate and of providing historical and archaeological evidence on the co-evolutionary relationships between human settlement and water availability in semi-arid environments. The link between governance and resilience is also addressed, though this is brief as there is little empirical evidence to support characterization of such dynamics.

Water resources include surface and groundwater, inland water, rivers, lakes and wetlands. These resources have attracted significant interest in the economics literature because of the diversity of values they possess and the challenges involved in their accurate estimation. In order to achieve efficient and equitable water resources management it is necessary for policymakers to have appropriate estimates of the values for water resources to be used in further analyses. In Chapter 5, Ekin Birol, Phoebe Koundouri and Yiannis Kountouris present the issues involved in the economic valuation of water resources. The authors first define the concept of the Total Economic Value (TEV) of a resource and list the plethora of values of sub-values accruing to water resources, following the distinction between use and non-use values. The methodologies employed by economists for the valuation of environmental goods and natural resources are presented in the next sections highlighting their respective strengths and weaknesses. These approaches can be broadly distinguished in revealed and stated preference methods. Revealed preference methods examine related market where the environmental good is traded and information derived from those markets is used to estimate the Willingness to Pay of the individual for the environmental good which represents her valuation for it. Stated preference or direct valuation methods are employed for the cases where no markets exist for the environmental good under consideration. These are survey-based methods that can also be used for the estimation of non-use values. While presenting environmental valuation methodologies the chapter also reports on their applications on water resources valuation in Mediterranean and Southern European Countries.

In Chapter 6, Phoebe Koundouri highlights an important challenge of moving from research to policy in water resources management by stressing the significance of choosing the appropriate discount rate when conducting Cost Benefit Analysis (CBA) for projects with extremely long horizons. The EU WFD explicitly mentions that the policy maker needs to choose the most appropriate 'measures' in order to achieve "good water quality" by 2015. The choice among different "measures" and policies for water management is facilitated by (CBA). To implement a CBA, one needs to choose a discount rate. If the chosen discount rate is wrong, then the results of the CBA are wrong and thus the chosen measures and management options are inefficient, non-equitable and unsustainable. Whereas the conventional view has always been that there is a unique social discount rate – the value of which has been disputed over 30 years or so of debate – new work suggests powerful reasons why the discount rate is not a single number, but a number that varies in a declining fashion with time. The chapter provides a non-technical review of the formal justifications supporting the use of Declining Discount Rates (DDR). It proceeds to illustrate the arguments for DDR by presenting a case study of applying different discounting schemes to a CBA for flood defences investment in Shrewsbury in the UK. In particular, two flat and four declining discounting rate schemes are applied and their policy implications regarding the adoption of the flood defences project are reported and compared. These results can aid in the design of sustainable and equitable policies for integrated water management, with Europe-wide implications. The message of this chapter is that policy makers need to understand that the constant discount rate that they continue to use in the implementation of the WFD is inappropriate.

Many research efforts have recently began exploring means and ways to tap into the yet unrealized support that models and the modelling process could offer to participatory river basin planning for IWRM. It is within this context that Chapter 7 originates, with the aim of building on recent research experiences to offer insights into future research needs in support of participatory planning for integrated water management. More specifically, the purpose Carlo Giupponi and Alessandra Sgobbi in this chapter is to illustrate how models, in the broad meaning of the term, could support the integration of political and social dimensions in IWRM. In order to put the present research into context, they analyse in more detail the role of Public Participation (PP) in natural resources management, while the following section explores the use of the terms "model" and "DSS" (Decisions Support Systems) within the IWRM paradigm. They also discuss the specific experience of the MULINO Project ("Multi-sectoral Integrated and Operational Decisional Support System for Sustainable Use of Water Resources at the Catchment Scale") for the implementation of the concepts of IWRM, with specific reference to the EU

Water Framework Directive and its implementation process, drawing some general lessons for what concerns public participation. The last section of the chapter presents some concluding remarks and insights for future research agenda for improving the effectiveness of participatory modelling, with focus on the problems typical of the Mediterranean Region.

Water management and development policy often neglect the importance of institutions and the synergies among past, present and planned policy interventions and the impacts of policy changes. Following this observation, in Chapter 8, Maria Saleth, Ariel Dinar and Susanne Neubert present a unified framework for taking into account the institutional impacts and development synergies in achieving meta development goals such as the Millennium Development Goals (MDG) and especially food security, while also presenting a methodology for testing their significance quantitatively. In their analysis, the authors build on the institutional ecology framework that allows basin institutions to be viewed under a given social, economic and physical context. This is combined with the institutional decomposition and analysis approach in order to identify and reveal the linkages between basin institutions and the adaptive instrumental evaluation approach that facilitates the collection of ex ante qualitative information on the institutions from stakeholders. Following this combined methodology the authors develop a representation of the linkages between development policies, institutional configurations impact pathways and the goal of food security and proceed to present a stylized system of equations capturing all interactions and links. The model is then applied in the case of Kala Oya Basin in Sri Lanka: a Three Stage Least Squares procedure is applied for the simultaneous estimation of 21 equations composed of 32 institutional, development intervention and impact variables using qualitative data from stakeholder surveys. The results reveal the importance of institutions in the generation and transmission of impacts and the synergies among various policy interventions.

The significant interlinkages between poverty and the state of the environment, including those between poverty and the shortage of water, are getting increased recognition, but little consideration is being given to the fact that protecting ecosystems directly or indirectly related to water is crucial to sustainable development. Fears that protecting the environment may hinder development are ungrounded since more and more evidence shows that environmental protection and development are mutually reinforcing. One of the most important health hazards, particularly for urban dwellers in developing countries, is extend contamination of water and food due to poor or non-existent sanitation systems and inadequate hygiene, compounded by unreliable and unsafe drinking water supply. The conservation of ecosystems that are water-related, directly or indirectly, should therefore be the basis of any strategy aimed at achieving poverty eradication on a sustainable basis, through, inter alia, the provision of reliable, sufficient and good-quality water. At the same time given that water is not a commercial product like any other, but rather a precious heritage efficient pricing is significant to act as an incentive for long-term sustainable water resources use. Firstly, water pricing will limit demand to efficient levels and secondly revenues will be generated which can ensure the maintenance of reticulation systems and the reliability of water supply. Chapter 9 by Ben Groom, Xiaoying Liu and Tim Swanson addresses these issues by examining alternative water pricing strategies that will ensure water cost recovery and identifying their impact on the socio-economic environment with emphasis to the poverty alleviation target. The study is based on data obtained from arid North East of China, and in particular Beijing, a region suffering serious water shortages. Authors analyse one pricing policy, the implementation of Increasing Block Tariffs (IBTs), which is often claimed to alleviate the equity and poverty concerns surrounding water pricing by subsidising "lifeline" levels of water consumption. Panel data were used to determine the exact (Hicksian) welfare impacts on low-income groups of a move from uniform full cost pricing to the proposed IBT regime. Results indicate that in the context of Beijing although IBTs are more equitable, since they do soften the blow to the poor, they are unlikely to circumvent the equity-efficiency trade-off completely.

Since water sustains all life, effective management of water resources demands a holistic approach, linking social and economic development with protection of natural ecosystems. The EU defines a holistic approach to water resources management as one that encompasses "environmentally-sound water management; food security especially for the poor; private sector involvement; reduction of subsidies; decentralization of decision-making to the lowest appropriate administrative level; user participation in services; institutional reform and regulatory frameworks; and cost recovery and pricing". In line with the Water Framework Directive, this book stresses the need for an Integrated Water Resources Management (IWRM) approach to balance the competing demands on water – domestic, municipal, agricultural, industrial and environmental – and promote conservation and sustainable use in an equitable fashion in many regions in the Mediterranean and the developing world. Research outcomes of the projects included in this book, highly demonstrate that effective and appropriate water management tools and decision-making practices are needed in order to complement integrated interventions for increasing the availability of supply and/or managing the growing demand of scarce water supplies. Further the book attempts to bridge the gap between ideas and actions endorsed at the research-oriented environmental debate, and their translation into policymaking structures and programs in developed and developing countries.

Part I
Results from the ARID Cluster and other European Research Projects

Chapter 2
Water Management on Mediterranean Islands: Pressure, Recommended Policy and Management Options

Antonia A. Donta, Manfred A. Lange and the MEDIS consortium

2.1 Introduction

Water is essential to life on our planet. Our very existence as well as our economic activities are totally dependent upon this precious resource – and on a global level, water is in many cases a limited resource. Even if Europe does not face generally shortages, Europe's water is far from satisfactory with regard to its quality and management. As the most sensitive area in Europe, the Mediterranean is particularly affected by this problem. Moreover, this situation on the Mediterranean islands is further aggravated because of their geographical isolation. Thus it is impossible to draw on more distant or diverse aquifers in general. Likewise, the threat of saline intrusion reduces the utilization of existing near shore aquifers in particular. Furthermore, the situation worsens during the driest part of the year between May and October when conflicting water demands and water availability are imbalanced. While the average precipitation within this period is very low the water demand is quite high. The water needs of the agricultural sector as the major user of water are very high and occur at the same period as the considerable water needs of the tourist sector. Therefore an increased water demand can be observed in summer. In addition, an inappropriate use of agrochemicals reduces the water quality and thus again water availability.

The European project MEDIS aims at finding solutions to this predicament, following a holistic, interdisciplinary approach, and at deriving recommendations, e.g., generic solutions, for the Mediterranean. Such solutions are also pertinent in the context of the European Water Framework Directive which came into force on 22 December 2000 and which aims at supporting all water-protecting in order to achieve or maintain a good status, thus fostering a sustainable water use to guarantee a long-term protection of available water resources.

For reaching the main goal of MEDIS, indicators for the natural environment (climate, hydrology, etc.), as well as for socio-economic conditions have been employed for the five islands Majorca, Corsica, Sicily, Crete and Cyprus. In general, the DPSIR methodology (Driving forces, Pressures, State, Impact, Responses) has been followed to identify sensitive conditions/hot spots and achieve a synthesis in order to derive the most appropriate recommendations for water management on

P. Koundouri (ed.), *Coping with Water Deficiency,* 11–44.

the Mediterranean islands and the wider Mediterranean area by consulting local stakeholders as a key of the approach. Local stakeholders play an essential role in MEDIS, firstly because they know about the particular needs of their location and thus possess the necessary expertise and experience in finding the best possible solution to water management. Secondly, recommendations will only be accepted by the population if these fulfil and are in accordance with their needs.

2.2 Methodology

The process applied here for deriving management options is based especially on the analytical framework known as DPSIR (Driving forces, Pressures, State, Impact, Responses). This framework developed by the United Nations is based on the Pressure-State-Response (PSR) framework that was first developed by the OECD in the late 1980s (Walmsley, 2002). According to that system analysis view, human activities (social and economic developments) exert pressure on and thus affect the environment as such. As a result the environment changes its state (health, resources availability, biodiversity), which might lead to impacts on human health, ecosystems and materials. As a consequence that may elicit a societal response, in return responding to the driving forces or the state or impacts directly through adaptation or curative action.

Consequently, this framework allows a comprehensive assessment by showing the chain of causal links, such as between environmental problems, their impacts and society's responses to them (cause–effect logic), in an integrated way by using indicators, even if the real world is far more complex than can be expressed in simple causal relations in system analysis. Nevertheless, from the policy point of view, there is a need for clear and specific information. The DPSIR framework is hence useful in describing relationships between the origins and consequences of environmental problems and can be considered as a tool for organizing environmental information and presenting causal connections between environmental indicators for decision makers.

Environmental indicators provide information about phenomena that are regarded typical of and essential to environmental quality. They enable a characterization of conditions, for example of water management conditions, thus both a comparison and, arising from that, a debate on environmental issues. Communication demands simplicity. Indicators are a useful means to simplify a complex reality by focusing on certain aspects regarded as relevant and on which data are available and are therefore useful tools to policy making (EEA, 1999, 2003a, b).

At present, the most indicator reports compile sets of physical, biological or chemical indicators. This paper considers indicators reflecting the natural environment and the human dimension in order to receive an overall characterization of the situation on the five islands under consideration. A comparison of the islands' water management situations will be carried out based on these indicators, and proposals will be given for a more sustainable and sound water management.

All data used within this paper originate from the island reports that have been compiled by the island teams, from deliverables within MEDIS and the internet.

2.3 Indicators Characterizing the Natural Environment

The main indicators applied to describe natural and environmental conditions are the aridity index, the water availability and exploitability, gross, person and sector water consumption. In addition water balance will be shown.

2.3.1 General Conditions

The five Mediterranean islands under consideration in the MEDIS project are Majorca, Corsica, Sicily, Crete and Cyprus. However much they have in common, there are also many differences: They all are located in the Mediterranean, thus subject to the Mediterranean climatic conditions, yet extend from the western to the eastern borders of a naturally divers Mediterranean area. A brief general characterization is given in Table 2.1: The islands' surfaces range from 3,640 km^2 for Majorca, 8,335 km^2 for Crete, 8,682 km^2 for Corsica, 9,251 km^2 for Cyprus (5,760 km^2 is the governmental controlled area) up to 28,000 km^2 for Sicily, the biggest island in the Mediterranean. The permanent population living on the islands, tourists visiting the islands and the population density show variations as well. The smallest island Majorca, receives the most tourists reaching twice to almost four times the amount of tourists visiting the other islands and has the second highest population density. Whereas Sicily has the highest population density and Corsica has the lowest.

Table 2.1 General conditions of the five Mediterranean islands: island surface, population, population density and tourists. (Data for Cyprus: CyStat Website, for Corsica, Sicily, Majorca and Crete: Donta et al, 2003a-e Island Reports.)

Islands	Majorca	Corsica	Sicily	Crete	Cyprus
Area [km^2]	3,640	8,682	28,000	8,335	9,251 5,760 g.c.
Total permanent population	609,150 (1996)	260,196 (1999)	5,076,700 (2001)	601,131 (2001)	689,565 (2001)
Population density inhab/km^2	167	30	198	72	120
Tourists total	76,107,380 (1996)	2,200,000 (2002)	3,720,000 (2001)	2,569,000 (2001)	2,700,000 (2000)
Tourists –equivalent population	210,373 (1996)	68,110 (2002)		140,767 (2001)	81,270 (2000)

g.c: under government control

2.3.2 *The Natural Environment*

2.3.2.1 Precipitation, Evapotranspiration, Aridity Index

The average annual precipitation on the islands is estimated to range between 467 mm/a in Cyprus and 927 mm/a on Crete (Fig. 2.1), thus reaching levels of those in northern European countries on average approximately 600–900 mm/a. As opposed to the similar precipitation levels of southern and northern European countries, the potential evapotranspiration of southern European countries reaches values from 1,000 mm/a to 1,600 mm/a which leads to a water shortage situation on the islands, specifically in summer. The mean annual actual evapotranspiration for example for Crete has been estimated to represent 75%–85% of the mean annual precipitation in low elevation areas (less than 300 m ASL) and 50%–70% in high elevation areas (Regional Government of Crete; Donta et al., 2003b). Due to this fact considering evapotranspiration is more applicable for stating climatic conditions as described through the aridity index (Fig. 2.2).

The aridity index defines the ratio of precipitated to evapotranspirated water. Obviously Cyprus is the only island under consideration in MEDIS that can be categorized as "semi-arid" implying a great water stress. Majorca, Crete and Sicily belong to the category "dry-sub-humid" inferring no notable natural/physical water availability difficulties. Moreover, Corsica with moist, sub-humid conditions would not show any water-stressed situation, when its natural conditions are taken into consideration.

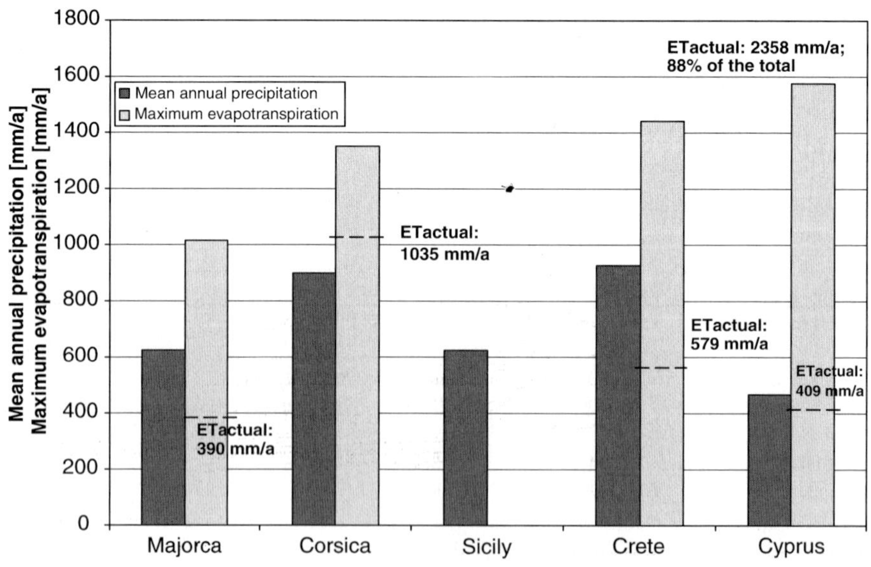

Fig. 2.1 Mean annual precipitation and evapotranspiration of the five islands under consideration in MEDIS

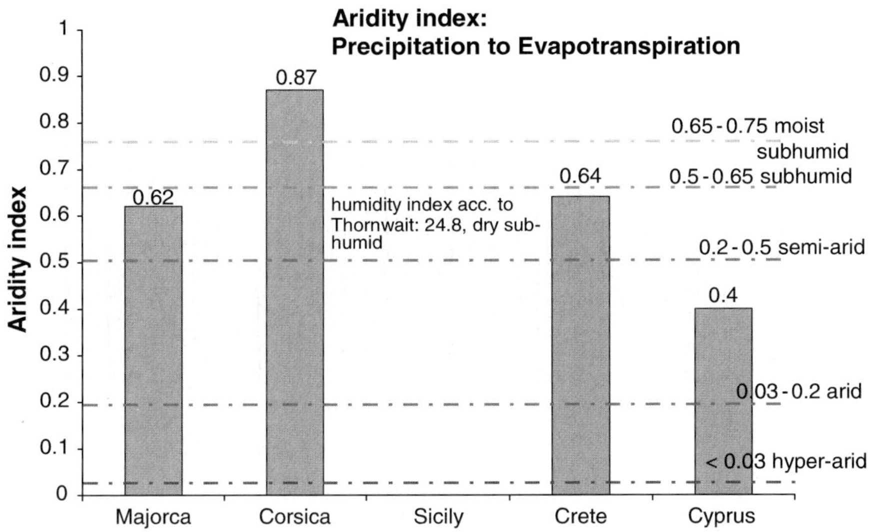

Fig. 2.2 Aridity index of the five islands under consideration in MEDIS

2.3.2.1.1 Inter-Annual Variations

The inter-annual variation of precipitation and, more specifically in light of their management, the seasonal and topographical variation of precipitation is of more importance to both the available water resources and reserves and for their management.

All islands show significant variations in their inter-annual precipitation. For example, the inter-annual variation in the prefecture of Rethymnon on Crete and Cyprus (Fig. 2.3 and 2.4) where rainy and wet as well as very dry and drought years or even periods can be observed.

2.3.2.1.2 Seasonal and Regional/Geographical Variations within a Year

With regard to the seasonal and topographical variations throughout the year all islands show similar characteristics. Rainfall is not uniformly distributed: 60%–80% of the precipitation is mainly concentrated in the winter months between September/ October and January sometimes expanding into April, whereas less than 10% of the precipitation falls during the summer season (Fig. 2.5 and 2.6). Precipitation during the water-intense period from April to October is seven times lower than from November to March showing an extremely stressed situation in this period for all islands (Fig. 2.6).

Yet Majorca shows a more even distribution in monthly precipitation ranging from 34 mm/month to 63 mm/month than Crete whose precipitation varies between 1 mm/month to 141 mm/month. Sicily and Cyprus take a middle position.

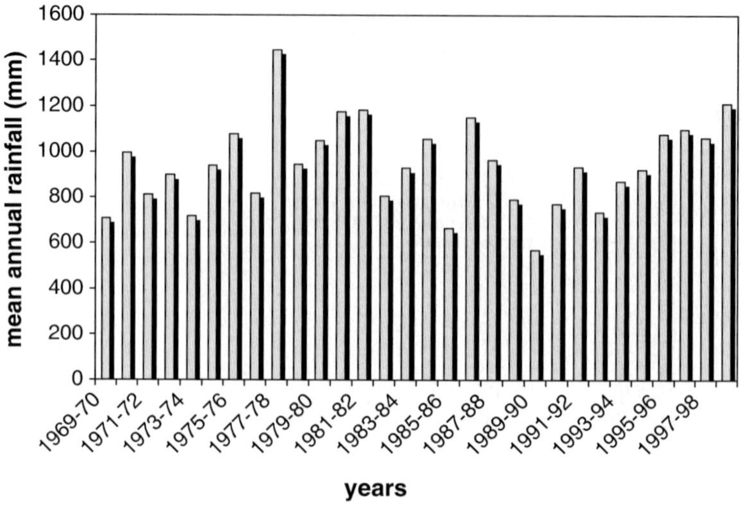

Fig. 2.3 Mean annual precipitation for the prefecture in Rethymnon, Crete

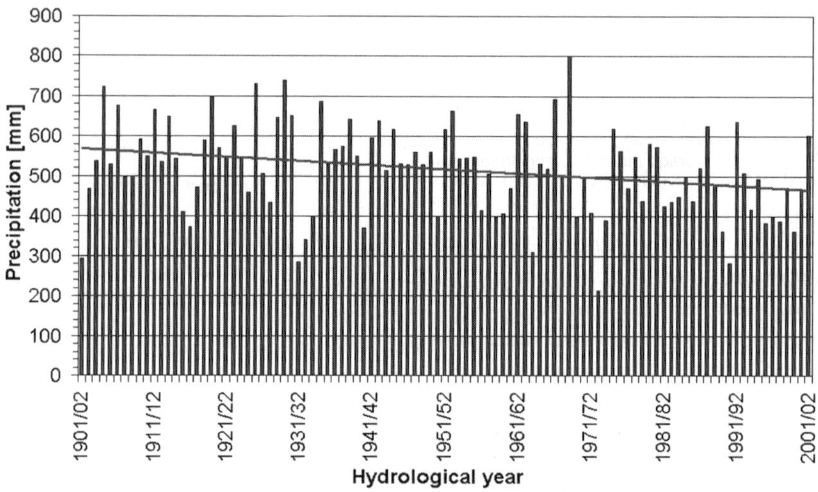

Fig. 2.4 Mean annual precipitation Cyprus wide: 1902–1997. The depicted data are taken from a station on Troodos (Amiantos) and from a station in the plains (Nicosia). (Geological Survey Study, 1998, Donta et al, 2003c)

Considering these climatic conditions the following policy options can be proposed for tackling the situation:

→ **Policy option category: Demand management**

- Promote water saving/conservation and storage measures in periods of high precipitation
- Realize rain water harvesting
- Reduce evaporation through covering open reservoirs, where feasible

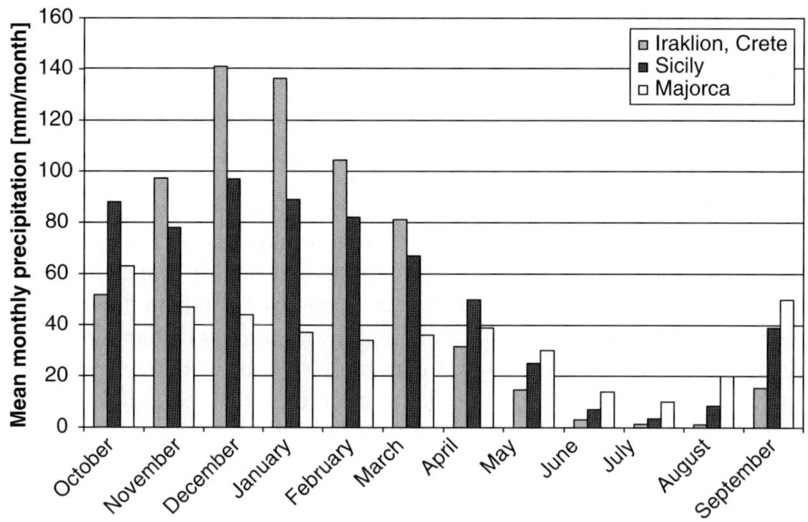

Fig. 2.5 Mean monthly rainfall pattern for the prefecture of Iraklion, Crete, Majorca (1972–1990) and Sicily (1965–1994) (From Crete: Regional Government of Crete, in Donta et al., 2003, Majorca: http://boreal.inm.es/wwc/html/dclimat/PALMA_.html, Sicily Regional Climatological Atlas, in Donta et al., 2003e)

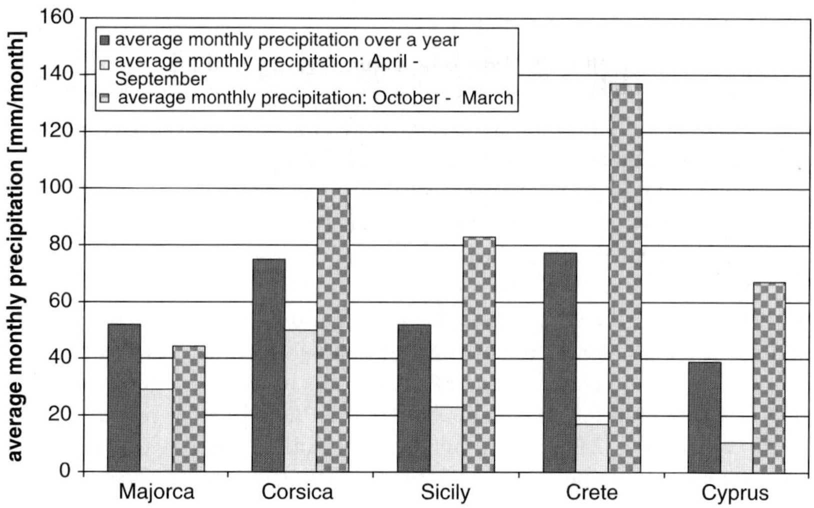

Fig. 2.6 Seasonal average monthly precipitation for the islands under consideration

Regarding the geographical rainfall distribution mountainous areas have a great deal of precipitation as compared to plains and coastal areas. Precipitation in Corsica, for example, ranges from 500 to 600 mm on the coastal periphery of the north of Corsica Cape and on low plains and valleys, to more than 1,500 mm on the mountainous areas of the central chain of Incudine to Cinto. Moreover, from

November to April there is a permanent snow-covering on the relief (above 1,400 m on northern slopes and 1,700 m on southern slopes) that could be potentially used as a water source after May. In Corsica, mountainous zones are less affected by climatic changes. Regions of the interior and those of the western frontage of the island are more affected by a rainfall deficit tendency than the regions of the Eastern frontage. Similar findings appear for Majorca. In mountainous areas the maximum can exceed 1,400 mm/a, whereas the minimum on the coast is lower than 300 mm/a. Cyprus' precipitation is in general very low and ranges from 300 mm/a, which characterizes desert climatic conditions, to approximately 700 mm/a. Thus Cyprus experiences a similar situation throughout the whole island in both its mountainous and plain areas. The decrease in rainfall is similar in plains and mountains. In Sicily the annual precipitation ranges from a maximum of approximately 1600 mm/a in the mountainous areas of the Etna massif to a minimum of about 300 mm/a in the southern coastal region. The annual rainfall for Crete ranges from 300 to 700 mm in the lower areas and along the coast (Ierapetra 312 mm, Iraklio 512 mm and Chania 665 mm) and from 700 to 1,000 mm/a in the plains of the mainland, while in the mountainous areas it reaches up to 2,000 mm (Chartzoulakis et al., 2001). Furthermore the southern regions of the islands are warmer than the northern ones, especially in Crete.

These (seasonal and geographical) variations and differences in water availability throughout the year make planning and managing water resources so difficult, especially in summer, leading to conflicts among user groups. Due to the prevailing stress brought about by reduced rainfalls and increased evapotranspiration in summer, where water demand is at its highest due to risen agricultural and tourist needs for water, islands face extreme water management difficulties. For the characterization of these types of conditions a significant indicator still needs to be developed, though these facts are the major pressure and driving force to be considered in recommending policy options.

Owing to the precipitation trend in the last years Cyprus has shown a severe decrease of mean annual rainfalls by 14% from 560 to 480 mm of the whole island over the last hundred years, Crete however, has not been affected by any decrease for the last 30 years. If the last three decades of rainfall in Cyprus are considered as well (Fig. 2.7), then no significant decline can be observed for its rainfall either. According to Rambaud (Raumbaud, 2003), Corsica has seen a significant decrease tendency of annual rainfall in general since 1984. Yet while the rainfall in Petreto Bichisano, Corsica (Fig. 2.8) shows a significant decline, in Ghisoni, Corsica (Fig. 2.9) no such decrease is found. Divided into decades, Corsica has been alternately subject to very rainy and drought-stricken years: 1971–1980 was a mainly "wet" decade, 1981–1990 can be considered as a "very dry" decade and 1991–2000 was primarily characterized by a succession of considerable surpluses and deficits. Thus, considering the last 30 years, it is difficult to state significant observations.

Regarding the precipitation tendency of these five islands of the Mediterranean Basin only Cyprus' precipitation development is in full agreement with general trends revealing gradual decreases in precipitation particularly for the Mediterranean Basin (Bolle, 2003, Xoplaki, 2002, Xoplaki et al., 2004).

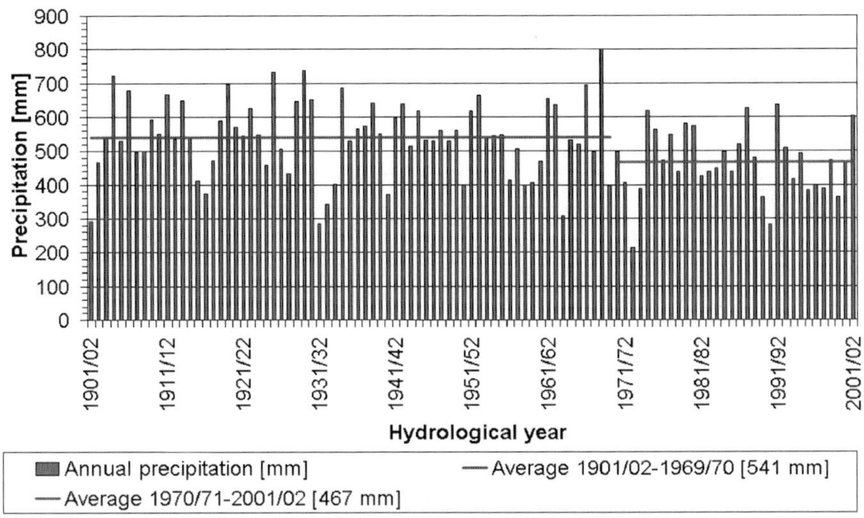

Fig. 2.7 Annual precipitation from 1901 to 2002 and average values for the years 1901 to 1970 (mean annual precipitation: 560 mm) and 1971 to 2002 (mean annual precipitation: 480 mm)

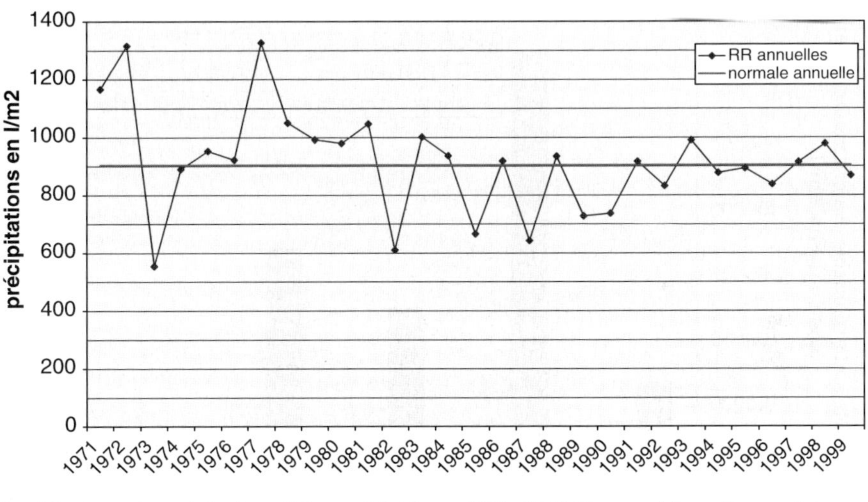

Fig. 2.8 Annual precipitation of Petreto Bichisano 1971–1999 (Source: Meteo France, in)

2.3.2.2 Available and Exploitable Water Resources

Figure 2.10 shows the water availability and the exploitable water resources per island area on each island. Available water resources include all natural renewable water resources (natural available water resources), thus surface and groundwater,

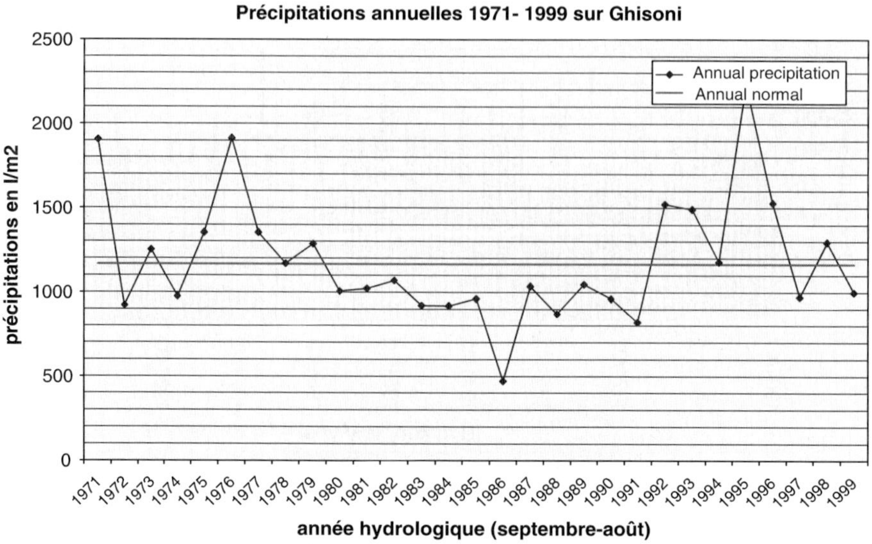

Fig. 2.9 Annual precipitation of Ghisoni 1971–1999 (Source: Meteo France, in Donta et al., 2003a)

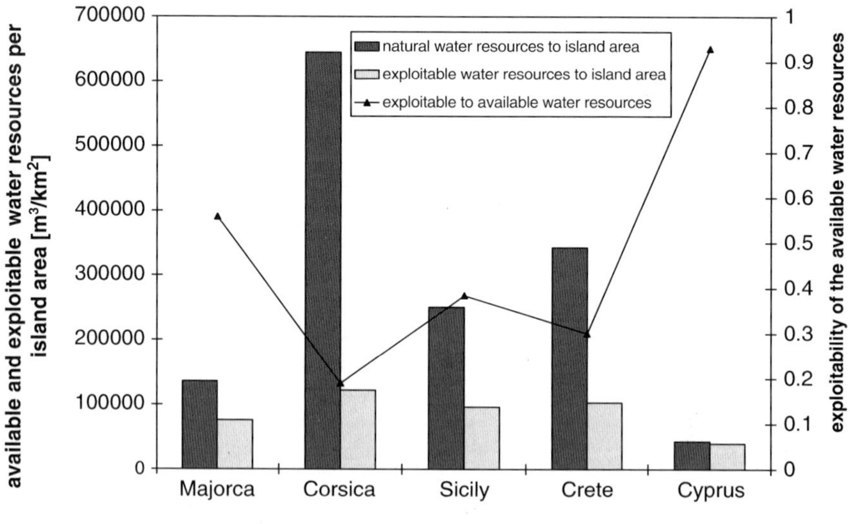

Fig. 2.10 Available and exploitable water resources on the five islands: Majorca, Corsica, Sicily, Crete and Cyprus

as well as water production resources, thus desalinated and wastewater treated water, usually used for irrigation in agriculture and the environment. It is obvious that the natural resources are not all exploitable. Exploitable water resources are the volume of water that could be technically and economically utilized without

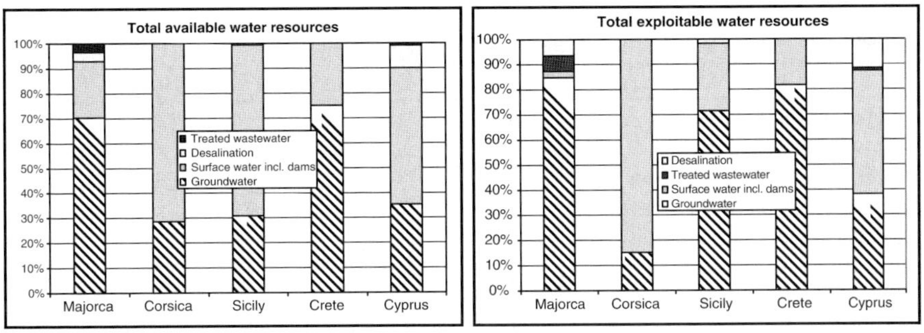

Fig. 2.11 Allocation of the total water resources available and exploitable in percentage

undesirable effects to ground or surface water potential[1]. Corsica shows the highest water availability per island area with 6,45,000 m^3/km^2, followed by Crete and Sicily. Yet, when regarding the exploitable water resources, the values of the islands lie closer to one another, range from 75,700 m^3/km^2 for Majorca up to 1,02,800 m^3/km^2 for Crete. Although Cyprus's available water resources are nearly completely exploitable (>90%), this amounts to 40,000 m^3/km^2, which is only half of Majorca's. For Corsica only its surface water is taken into consideration because there is no knowledge of the exploitability of its groundwater resources. Available and/ or exploitable water resources per area provide surface-oriented information on water availability of a region, enabling a comparison of areas or catchments varying in size.

Regarding the sources that constitute the available and exploitable water resources, groundwater is the main available water resource (70–75%) for Majorca and Crete (Fig. 2.11). Crete's availability of surface water and dams is 25%, Majorca's amounts to 22.5%. Both Majorca and Crete consume between 80% (Majorca) and 90–95% (Crete) of its exploitable groundwater. Cyprus', Sicily's and Corsica's groundwater resource amount to 35%, 31% and 29%, surface water including dams to 55%, 69% and 71% accordingly. The available surface water source for Cyprus is in fact dams, which explains the high exploitability of its "natural" water resources. Majorca's, Sicily's and Crete's and, to a lesser extent, Cyprus' main exploitable source is groundwater. Although groundwater availability for Sicily amounts to only 30%, 78% of it is exploited, which might generate problems in the future, depending on the ratio of renewed to abstracted groundwater (ratio of renewable water resources to water consumption is 0.53) and the intensity of groundwater use which is 78%. Corsica's aquifers are difficult to investigate, thus its groundwater availability and exploitability are not known. Therefore, surface water is the most utilized resource in Corsica (70%) which is also the main source of consumption (70%). Cyprus' available resources rather comply with those of Majorca, though

[1] Definition from the international glossary for Hydrology, UNESCO: Exploitable water resources is the available water, or water capable of being made available for use in sufficient quantity and quality at a location and over a period of time appropriate for an identifiable demand.

Table 2.2 Water balance based on yearly recharge and abstraction values. (Data from Donta et al., 2003a-e, Island Reports, for Cyprus: WDF&FAO, 2002.)

	Majorca	Corsica	Sicily	Crete	Cyprus
Precipitation (mean, last 30 years period [mm/a]	625	900	625	927	467
Renewable groundwater [10^6 m^3/a]	374.1	700	1,994.4	2,120[a]	208
Surface water recharge [10^6 m^3/a]	120	4,900	4,449.5		121
Renewable waters [10^6 m^3/a]	494	5,600	6,443.9	2,120	329
Total water available (incl. man-made) [10^6 m^3] /a	531.7	5,600		2,860	397.8
Water exploitable (incl. man-made) [10^6 m^3/a]	294.9	1, 060		857	370
Water abstraction [10^6 m^3/a]	233.3	86	1,874.8	422	229.4

[a] includes surface water

Cyprus has fewer groundwater resources amounting to 21%, of which only 40% are exploitable, with 45% of that source already being used. Cyprus' groundwater recharge only amounts to 208 $10^6 m^3$/a, which is only about half of that of Majorca and almost the annually consumed water resources of Cyprus itself (s. Table 2.2). This indicates once more that Cyprus faces severe water stress and depends not only on its surface water (dams) but more importantly on non-conventional sources like desalination and wastewater reuse.

2.3.2.3 Origin of Water Used

The main water source used on Crete, Majorca and Sicily is groundwater, with 93.8%, 82.8% and 77.7% respectively. Cyprus and Corsica use groundwater to a lesser extent, i.e., 48% and 29.5% respectively. Corsica's main source is surface water (70.5%), while Cyprus has a more even distribution in its various water sources and uses 37.5% of its surface water. Corsica is the only island whose main water source is surface water. Majorca and Cyprus are the only islands that use treated wastewater, even though the proportion is rather small, with about 5–7% for Majorca and 1–2% for Cyprus. Majorca, Cyprus and Sicily use desalinated water as well; this source is also used in small amounts: 12% for Cyprus, 3–7% for Majorca and 1.2% for Sicily.

For the domestic water use, groundwater is the main source for Crete with a full 100%, for Sicily it amounts to 86% and for Majorca to 75%. Corsica's main source is groundwater as well but to a lesser extent, yet still reaching 60%. Cyprus uses 44% desalinated water and 31.5% water from dams in the domestic sector. Corsica's second source is dam water as well (40%). Since 2000 desalinated water has been used in Majorca which amounts to 18.4 $10^6 m^3$ (~3–7%).

Interestingly, groundwater is also the main source of irrigation on Crete and Majorca (92.3%, 90.6% respectively), likewise in Sicily with 68%. Although groundwater is not used as the only and main source in Cyprus, 53% is still a high proportion considering the low groundwater availability and renewability. Only Corsica uses dam water as its main source of irrigation (100%). Cyprus uses 45% dam water as well for irrigation purposes. Treated wastewater or other low quality waters, however, are not used to the extent in which they could with regard to irrigation. Only Majorca utilizes treated wastewater to a mentionable quantity of up to 7%, even though this is still too low.

Proposed suggestions in this regard are:

→Policy option category: Supply enhancement

- Use additional sources for irrigation purposes such as wastewater, treated or not, brakish water, grey water or other waters of low quality
- Instal home recycling systems and use grey water in the touristic sector as well
- Use waters of low quality (brakish, wastewater etc.) to replenish the aquifers and enhance available water resources
- Use more surface water instead of groundwater particularly for irrigation purposes but also for potable water to avoid further sea water intrusion and overexploitation of the aquifers in order to give the groundwater more time to replenish
- Use desalinated water and at the same time searching for methods that produce desalinated water at an affordable price, use of desalinated water in tourism

→Policy option category: Institutional policies/Regulations

- Calculate and consider rather exploitable values for planning than the natural available water resources; water resources exploitability range from 19% (Corsica) to 38% (Sicily)

2.3.2.4 Water Balance

Sustainable water availability depends on parameters such as precipitation, evapotransformation, infiltration, distribution and conveyance water losses that do not return into the water cycle, as well as water flows, i.e., losses, to the sea, specifically for islands. These are substantial factors in water balance and the sustainable use of water. The volume of water that is being abstracted compared to the available/exploitable water resources constitutes an additional factor in water balance.

The water balance of almost each of the islands is positive, though that of Majorca is very tight, and especially that of Cyprus depends predominately on the yearly precipitation but also on the additional sources of treated wastewater and desalinated water. With regard to Cyprus, there have been years during which the water balance was positive and years when it was negative, showing the extremely crucial situation Cyprus is in (Fig. 2.12, data from 2000). Both Majorca and Cyprus face the greatest water stress conditions regarding their available water resources which make their management challenging. For example, in 2000 the water needs for Cyprus in all sectors were 265.9 $10^6 m^3$, only 211.5 $10^6 m^3$ could be supplied, thus a water

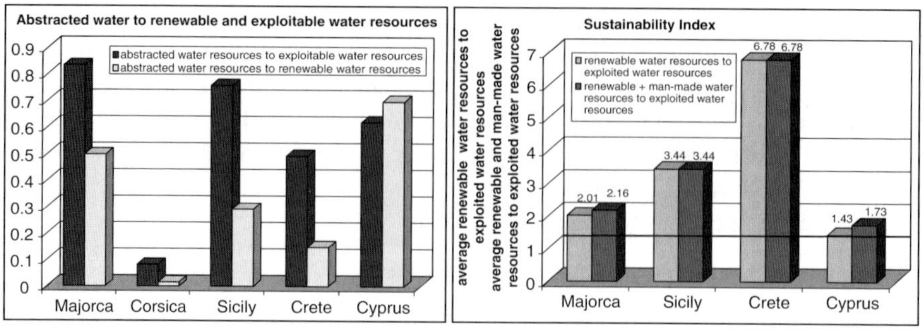

Fig. 2.12 Indicators for identifying sustainability of water use; index of water scarcity: water abstracted to exploitable water resources, renewable to exploitable water resources

shortage of 20% existed. Each of the sectors experienced a shortage, the shortage allocation was as such: the domestic sector including industry and tourism suffered a shortage of 23.5% from surface water, agriculture (irrigation) had a shortage of 45.6% from surface water and all sectors together experienced a groundwater shortage of 10% (after FAO&WDD, 2002). Concurrently there were 15% water losses in the agricultural sector and 20% losses in the domestic sector, which adds up to 15% water losses. Thus an additional constraint to the natural availability of water is the water losses caused within the conveyance and distribution system, which all islands suffer from, and which reaches up to 40%.

→Policy option category: Demand management

- Improve infrastructure to reduce water losses in the conveyance and distribution system for domestic and agricultural purposes, even though the lost water partly returns into the water cycle
- Reduce any losses to the sea from overland flow, through flow (sub-surface flow) and groundwater flow for example by developing wetlands or increasing capacity of storage/dams, by paying greatest need to the natural environment

These circumstances have already impelled Cyprus and Majorca to use additional water sources such as desalinated water and treated wastewater in order to combat this situation, as mentioned above and shown in Fig. 2.11. Cyprus started with its first desalination plant in 1999, Majorca in 1999. Cyprus began using treated wastewater for irrigation in 1995. In future, these additional water sources will ensure the supply of water in Cyprus when all works have been completed by 2015, as stated by the Water Division Department of Cyprus.

The sustainability index expresses the sustainable use of water and is given by the rate of groundwater recharge to the average water exploitation. If the water usage is sustainable, the ratio is greater than one (Veraart et al., 2004). Yet this index does not take into account the water needs necessary to maintain the ecological integrity of present ecosystems. The sustainability index presented here, however, includes all kinds of renewable water resources and thus water which contributes to preserve natural components and the environment.

Accordingly, water management in Cyprus is not sustainable, and as was previously stated, depends on the annual precipitation. When considering the renewability of water resources, Crete's sustainability index is high. This situation results from a high volume of renewable water resources and a relatively low water abstraction compared to the overall natural water resources. Yet not all renewable water can be made available for use on Crete, because about 80% of the aquifers are associated with deep karstic counterparts and both the technical and economic investments for expanding the exploitability of further water resources would outweigh any profit to be gained from further water exploitation. According to the Regional Governor of Crete overexploited aquifers already exist, for example the situation in the Messara valley where a drop in the groundwater level has been observed. This is mainly due to the over-pumping of aquifers for irrigation purposes. Similar situations have also been detected in other valleys used for agriculture.

While the yearly precipitation of each island except Cyprus reaches precipitation heights in average equal to those of northern Europe, as was aforementioned, their water availability and exploitability is comparatively rather low.
Further strategies to combat this situation can be proposed:

→Policy option category: Demand management

- Reduce water requirements for the various sectors
- Increase artificial recharge of groundwater by using water of low quality
- Improve conservation measures and water saving systems
- Make use of water recycling in industry and for domestic purposes
- Reduce unaccounted water by adding meters, enhancing extraction control and checking meters.

→Policy option category: Supply management

- Use wastewater (treated or not) for agricultural use, parks, golf places, private gardens (water recycling from septic tanks)
- Use brakish water in tourism, for example in swimming pools
- Use saline water for recreation, swimming pools, for salt-tolerant crops
- Use grey water for park irrigation, or irrigating specific crops in agriculture
- Utilize desalinated water by flexible units for seasonal demand. Even though desalination is an expensive technology it can help to tackle limited amounts of water.

2.3.2.5 Water Consumption

Water Used by Sectors - Yearly Water Consumption

The absolute volume of water consumed on each island differs extremely (Fig. 2.13) with Sicily using the most, 1,874 10^6 m^3 a year and Corsica the least, 61 10^6 m^3 a year. For reasons of comparison, it is advantageous to allocate the consumed water to the island area which gives more reasonable information on the water-utilization situation on the islands (Table 2.3).

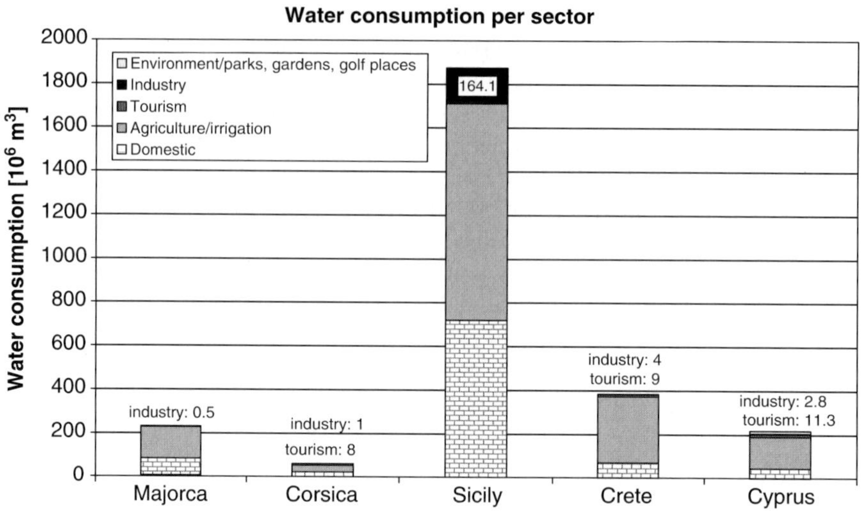

Fig. 2.13 Water consumption distributed per sector for Majorca, Corsica, Sicily, Crete and Cyprus

Table 2.3 Water consumption per area and year (data from Donta et al., 20003a-e Island Reports, for Cyprus: WDF&FAO, 2002)

	Majorca	Corsica	Sicily	Crete	Cyprus
Water consumption [10^6 m^3/a]	233.3	61	1,874.8	371.8	224.3
Area [km^2]	3,640	8,682	28,000	8,335	9,251 (Cy -wide) 5,760 (gov. contr. area)
Consumed water per area [10^3 m^3/km^2]	64.09	7.03	72.29	44.6	39

The ratios are: 7.03 10^3 $m^3/km^{2a}a$ for Corsica, 23.3 10^3 $m^3/km^{2a}a$ for Cyprus, 44.6 10^3 $m^3/km^{2a}a$ for Crete, 64.1 10^3 $m^3/km^{2a}a$ for Majorca and 72.3 10^3 $m^3/km^{2a}a$ for Sicily, showing that Sicily in fact uses more water per unit area than all the other islands, followed by Majorca, with the highest tourist population all year long and Crete with the highest agricultural production. Corsica has by far the lowest water use per area which is in accordance with its low water consumption and all indicator values presented so far. These ratios depend on the building density, the population density and the amount and size of irrigated areas used for agriculture (see below). It is remarkable that Majorca's water consumption per unit area is high. Therefore the question arises as to whether the high utilization generates from a great tourism consumption or a high water usage in agriculture. Consequently it is essential to study the water consumption in more detail.

If the water consumption per sector is compared in turns of percentage, it can be clearly seen that agriculture consumes the most water on all islands, ranging between 60% even on the main tourist island Majorca, to 80% on the predominately agricultural island of Crete (Fig. 2.14). This illustrates one reason why Majorca's

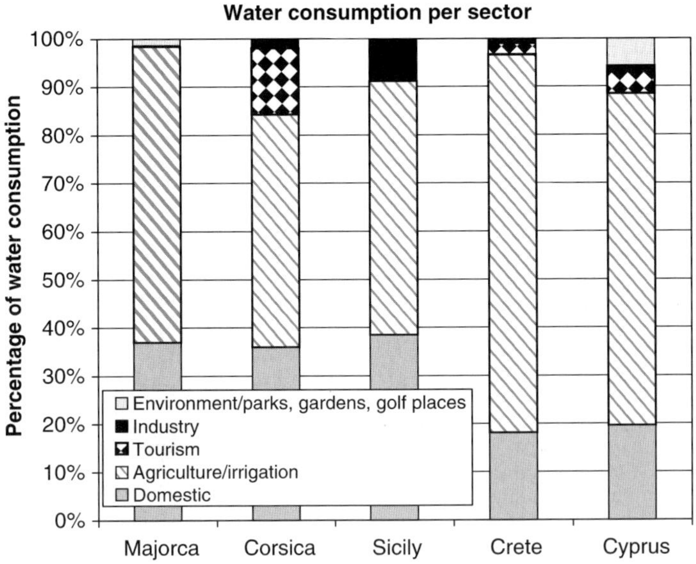

Fig. 2.14 Distribution in percentage of water consumption on the islands of Majorca, Corsica, Sicily, Crete and Cyprus in each sector

water consumption per unit area is so high, higher than that of Crete. In addition to its high water use in the tourism sector, Majorca uses 60% of its water in agriculture. The relation of the tourism water consumption to the gross water consumption ranges between 13% for Corsica, 5% for Cyprus and 2.4% for Crete. As a part of the domestic water use Cyprus has a 26.5, Crete a 13 and Corsica a 36.4 percent proportion of tourism water consumption within the domestic sector.

Daily Water Consumption

Regarding the daily per capita water consumption, the gross consumption, including the agricultural water usage is the highest for Crete, followed by Sicily, Majorca, Cyprus and Corsica. By comparing the domestic water use alone, which is given for Majorca and Cyprus, the daily water consumption varies between 150 to 465l/capita[a]day, whereas tourist cities use double to three times (up to 500 l/capita and day) as much as non-tourist cities (about 150l/capita and day), even during non-tourist periods. Considering these records, the contribution of the tourism water consumption to the overall water consumption seems to be of greater importance than is currently given, especially for the two islands Majorca and Cyprus which suffer from water shortage.

It is also remarkable that, when water consumption is regarded in its relation to the area, Sicily shows the highest allocation, yet when the consumption per capita and per day is looked at then Crete has the highest per capita water consumption.

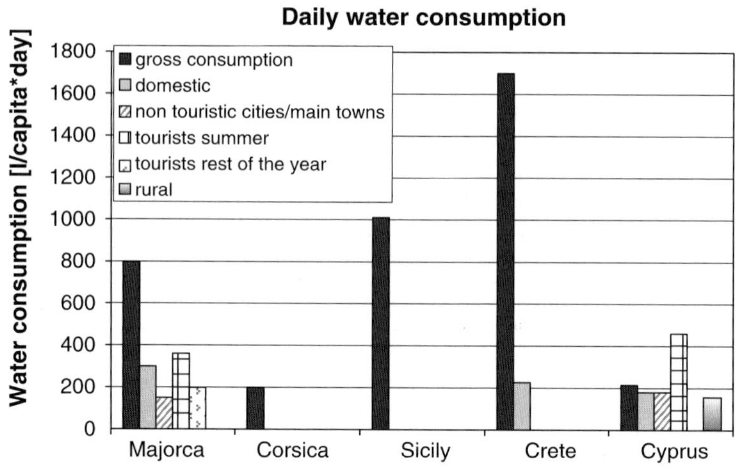

Fig. 2.15 Daily water consumption

This leads to the statement that Sicily's higher water consumption is due to the high population density, whereas Crete's high consumption is due to a high water use for irrigation. Even though the tourism sector comprises only a small and lesser amount in water usage as compared to the agricultural sector, measures have to be taken in order to minimize water misuse and wastage (Fig. 2.15).

→Policy option category: Demand Management

- Stimulate the use of water saving techniques in households, tourism facilities, industry and agricultural sector through *technical devices*
- Encourage the use of improved irrigation systems (drip irrigation) and irrigation scheduling techniques that are water efficient by setting standards, using expert systems, supplying an infrastructure and know-how in farming methods
- Provide assistance for investments in modern technologies

→Policy option category: Institutional policies

- Economic policy: reduce water consumption by introducing a water pricing system which recovers part or all costs by a two part tariff, a volumetric rate tariff, differentiated among the different user groups or other appropriate pricing models, yet transparent and reasonable for all consumer groups
- Provide *incentives* for reduced water consumption, for example by using water-saving techniques for households, hoteliers, farmers and so on
- Provide subsidies, for example for using grey water for toilets and garden irrigation
- Reduce water consumption by setting new standards in all sectors using water

→Policy option category: Development policies

- Examine the regional development policy and priorities given
- Raise environmental concerns, through legislation and information

2.4 Agricultural Indicators

To verify the impact of agriculture on water usage and management, a number of significant agricultural indicators will be introduced. The area which is used for agriculture in Sicily is 15,640 km^2, amounting to 60.9% of the island's surface (Fig. 2.16). This is the biggest share compared to the other islands implying a high water usage, however depending on the area which is irrigated as well as the crop's water requirements. On Majorca an area of 2,205 km^2 is cultivated, that with its 60.6% (PHB, 1999) of the island's total surface comes close to that of Sicily. Crete's area used for agriculture is 3,223 km^2 and Cyprus' 1,432 km^2, their share is 38.7% and 15.4% accordingly. Corsica utilizes 1,559 km^2 for agriculture which is the smallest share as compared to the island's surface.

It is remarkable that Majorca, as an overall tourist island, has a 60.8% share of agricultural land as related to its entire island's surface, Crete reaches a share of only 38.7%, by having the second highest gross water consumption after Sicily, which corresponds to the agricultural water use. These facts call for the more appropriate indicators which describe the repartition of the area being irrigated as compared to the area used for agriculture and the entire island area. These indicators provide information on how big the area requiring water is, related to the area which is utilized in agriculture and the domestic area. In addition, if this information is linked to the amount of water consumed and the crops cultivated, appropriate assessment of the water management in agriculture is possible. As can be seen in Fig. 2.17, the proportion of irrigated to cultivated agricultural land is 33.5% in Crete, which is comparatively higher than on the other islands. Also the share of the irrigated area to the island's surface is higher for Crete than for the other islands (Fig. 2.18). Both figures agree with the high water consumption in agriculture for

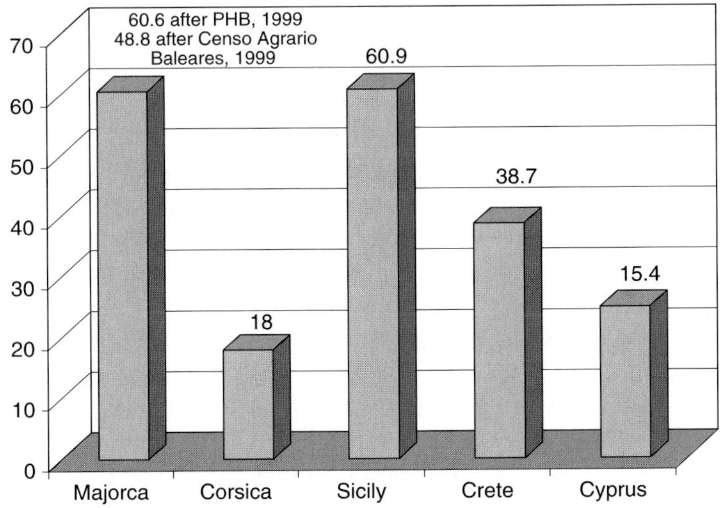

Fig. 2.16 Share of agricultural land to the total island's surface

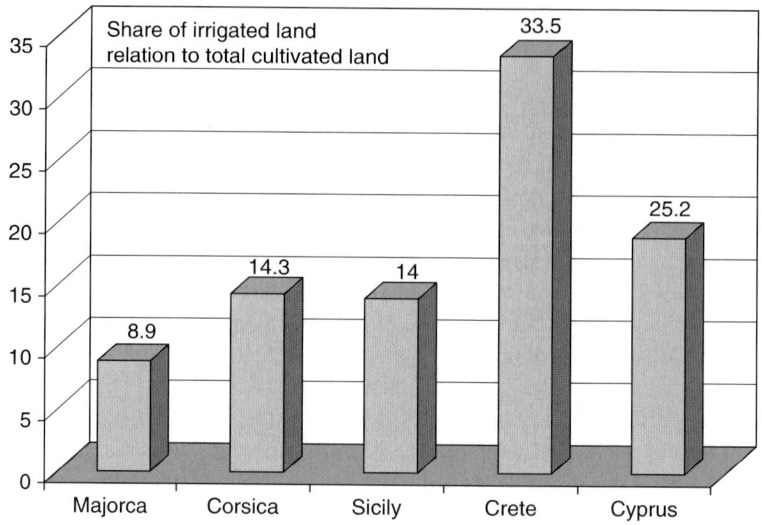

Fig. 2.17 Share of irrigated land to the total cultivated land

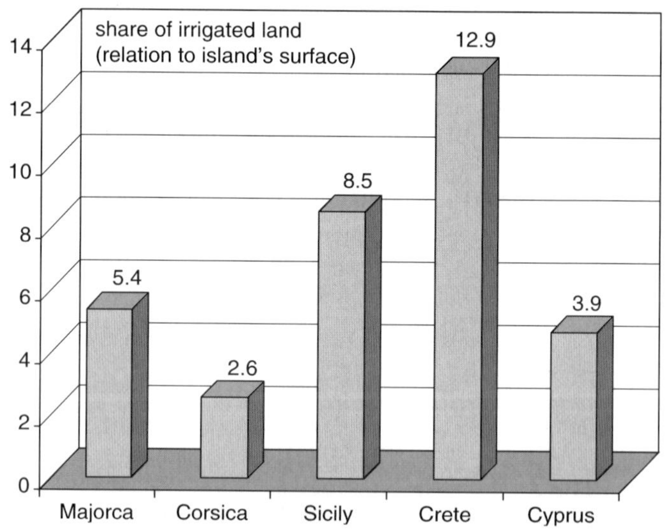

Fig. 2.18 Share of irrigated land to the total Island's surface

Crete. Cyprus' irrigated surface follows with a share of 25.2% for the cultivated land and 3.9% as part of the island's surface. Majorca's share of irrigated area within the cultivated area is the smallest, yet reaching 5.4% when related to the islands surface which lies above the accordant share for Cyprus and Corsica.

Figure 2.19 clearly shows that Majorca consumes the highest volume of water per irrigated area. Given this data, it is obvious that Majorca cultivates water intensive

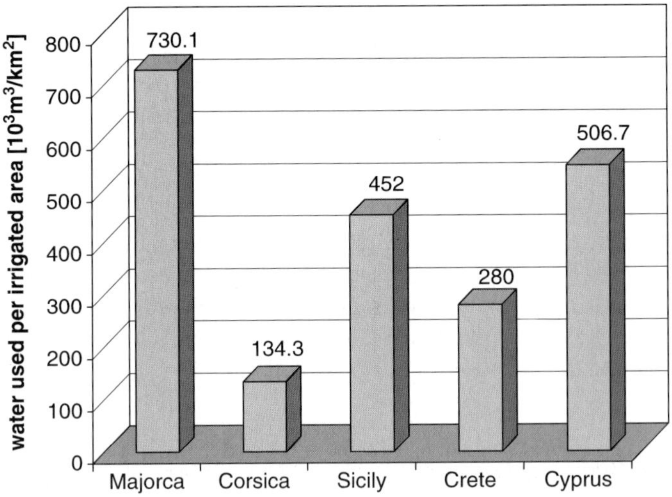

Fig. 2.19 Water consumption per unit irrigated area

crops or pursues an ineffective agricultural praxis, followed by Cyprus and Sicily. Yet, if the crops' water requirements are taken into account as well, the following ratios can be given: Cyprus has a ratio of water used for irrigation to crops' water needs (requirements) of 0.73 and if the water losses in irrigation of about 20% are excluded than the ratio falls to 0.62. The ratio for Crete is 0.66 and for Majorca 1. Accordingly, Majorca irrigates its crops in compliance to their needs, thus the low amount of water used per surface irrigated area which are given for Cyprus and Crete compared to that of Majorca is only due to the fact that these two islands fail to meet the crop water requirements. Undoubtedly, Cyprus is unable to cope with the situation due to a factual shortage of water resources. In 2000 water shortages constituted about 20% in all sectors, i.e., a 45.6% shortage in agriculture, 23.5 in domestic (tourism and industry included) (FAO & WDD, 2002). Thereby 15–20% were losses, in the domestic and the agricultural sector.

This situation regarding water consumption for agricultural use requires further investigation into the crops cultivated and their water requirements. Therefore a brief presentation of the main crop types and cultivars grown on the islands will follow. The description will follow the order given by the magnitude of water consumption per unit irrigated area.

2.4.1 Majorca

The main crops cultivated in Majorca and the area which they occupy are 54% row crops (grain cereals; 38.5% of the total cultivated area) followed by tree crops, nut trees (42%), almonds the most (23.8% of the total cultivated area) and forage crops 14.6% of the total cultivated area.

The major irrigated crops on Majorca and the area they occupy are shown in Table 2.4. Table 2.4 indicates that fodder crops (51.2%), vegetables (19.7%), tuber (2nd harvest; 10.6%) and citrus trees (10.2) occupy the largest part of the irrigated land.

On Majorca different sources for agricultural data are available, e.g. crop cultivation and water consumption by each crop (PHB 1999, Censo Agrari Conseleria d'Agricultura I Pesca, Conselleria d'Economia, Comerc I Industria, Govern Balear, 1999; DGOH, 1994, in agricultural report, 2004). Nevertheless each of these sources indicates that fodder/forage (60.8%) cultivation is the main consumer of irrigation water followed by vegetables (17.5%), potatoes (9.4%) and citrus trees (8.3%). Although the demand for irrigation water for pulses is high ($60\,m^3/km^2$) they occupy only a small part of irrigated land ($1.34\,km^2$) and are therefore not of great relevance to the increased water consumption.

2.4.2 *Cyprus*

The main crop categories in Cyprus are 55% tree crops, 38% vegetables, 5% vines and 2% row crops; the main cultivars (Table 2.5) are temporary crops which occupy 65% of the total cultivated area. Temporary crops include open field vegetables, greenhouse vegetables, flowers, potatoes and fodder.

In Cyprus the main irrigated crops are permanent crops such as citrus, deciduous, olives, bananas, table grapes and avocados. The annual irrigated crops include open field and greenhouse vegetables, flowers, potatoes and fodder. The deciduous crops include apples, peaches, cherries, pears, plums, figs, walnuts, pecan nuts and pomegranates. The main crops of the rain-fed cultivated land are cereals with a

Table 2.4 Main irrigated crops on Majorca (After PHB, 1999, in: agricultural report, Crete team, 2004.)

Crops	Surface km^2	% of irrigated land
Cereals	7.16	3.7
Fruit tree	5.26	2.7
Citrus tree	19.99	10.2
Vegetable	38.52	19.7
Tuber (2nd harvest)	20.62	10.6
Fodder	100.02	51.2
Pulse	1.34	0.7
Industrial crops	2.27	1.16

Table 2.5 Main cultivars in Cyprus

	Area [km2]	%
Temporary crops	934	65
Permanent crops	418	29
Fallow land	80	5.5

Table 2.6 Distribution of irrigation water demand by crop in Cyprus (WDD&FAO, 2002)

Category		Crop	Irrigated area [km^2]	Water applied [Mm3/a]	Share of total irrigation water
Tree crops		Citrus	70.84	55.81	32%
Tree crops		Deciduous	24.8	19.18	11%
Tree crops		Olives	19.85	8.72	5%
Vine		Table Grapes	20.07	5.23	3%
Bananas	Permanent Crops	Bananas	2.91	3.49	2%
Others		Olives, Flowers, Greenhouses, Almonds	14.0	10.46	6%
Vegetables		Green houses	3.21	3.49	2%
Vegetables		Open field vegetables	42.7	39.24	22.5%
Vegetables		Potatoes	8.64	16.57	9.5%
Arable land		Fodder (Alfalfa)		12.21	7%
	Annual Crops	Total water requirements		174.4	41%
		Real water irrigated (-15% losses)		108.8	

small portion (4.5%) of fallow and grazing land. Citrus crops take up the largest irrigated area (70.84 km^2), followed by vegetables (64.2 km^2), potatoes (42.7 km^2), deciduous crops (24.8 km^2) and vine (table grapes; 20 km^2). Although the irrigation requirements are the highest for lucern (1,350 mm/a), bananas (1,252 mm/a), pecan nuts (995 mm/a) and citrus and avocados (800 mm/a), the greater amount of water is irrigated to citrus trees and open field vegetables (Table 2.6).

The data of this table indicate that the greater part of irrigation water is consumed by permanent crops with citrus possessing the bigger share. Among the annual crops, the open field vegetables and potatoes are the main consumers of irrigated water whereas vegetables and flowers grown in greenhouses consume only 3.49 10^6 m^3 water which is 2% of the total irrigation water.

2.4.3 *Sicily*

The main crop types in Sicily are row crops (cereals) 58%, with wheat being the main cultivar (40.5%), followed by forage crops, vineyards, and olives.

2.4.4 *Crete*

The main crop categories on Crete are tree crops 68%, specifically olives being the main cultivars. Olives occupy 60% of the total cultivated area while vines account for 12% of the total cultivated area. On Crete the main irrigated crops are vegetables;

91–93% of the cultivated vegetables are irrigated, followed by vines with 45%. Also 36.3% of the fruit trees and 34% of the cultivated row crops are irrigated. The main crops of the rain-fed cultivated land are cereals with a small part of fallow and grazing land (4.5%).

Corsica

The main crop type and irrigated crop in Corsica are row crops, especially fodder crops: 67% of the total agricultural surface, including natural meadows and routes. Vines occupy 13.5% of the total agricultural surface, followed by citrus and olive trees. The share of irrigated surface to the total irrigated area per crop type amounts to 63% fodder, 11% tree crops (kiwi and other orchards), 10% citrus trees, 4.8% vine, 4.6% cereals, including corn and 2.4% fresh vegetables (AGRESTE census 2000 & AGRESTE Corse 2002, in. agricultural report, 2004).

Summarizing, the main irrigated crop types for each island are: fodder (60%) and vegetables (18%) on Majorca, citrus trees (32%) and open-field vegetables (22.5%) on Cyprus, fruit trees, specifically olives (68%) and vines (12%) on Crete, row crops (cereals) 58%, and forage/fodder crops, vineyards and olives and fodder (63%) and fruit trees (citrus 11%) for Corsica.

Crop water needs are influenced by climatic factors: sunshine, temperature, humidity and wind speed. The irrigation water needs are determined by the total water need of the various crops and the amount of rain water which is available to the crops. The water needs are determined by the daily water needs as well as the duration of the total growing season and the time of the year when a crop is grown. Given all these factors and according to a number of indicative values of crop water needs (Brouwer and Heibloem, 1986) alfalfa (fodder), bananas, citrus trees, tomatoes and various open-field vegetables are crops with high water requirements ranging between 600 to 2,200 mm/total growing period. Comparing the crops grown on the islands and the crops with high water needs, it is obvious that the majority of the cultivated crops belong to this category.

Another point influencing the amount and frequency of irrigated water is the sensitivity of the cultivated crops to drought conditions. Bananas and potatoes for example which are grown on Cyprus (potatoes are grown on Majorca and Crete as well) are very sensitive to droughts and therefore need special care and sufficient water supplies.

Given that agriculture is a main water consumer, the following suggestions could provide remedy in water usage for irrigation:

→ **Policy option category: Demand management**

- Examine the current agricultural practices/methods
- Modify and reconsider the current cropping patterns in favour of crops with fewer water requirements and less sensitivity to droughts, rain-fed crops, or water needs in winter
- Reflect on curtailing the irrigated area
- Encourage meetings aiming at conflict resolutions

→ Policy option category: Supply enhancement

- Use additional and unconventional water sources, use low quality water
- Set up a water bank (possibly directed by the authorities)

→ Policy option category: Institutional policies/Capacity building

- Inform farmers on the optimum weekly irrigation dose, by developing know-how
- Provide assistance in capacity building of farmers through teaching (courses, workshops, experimental exercises in the field)
- Encourage cooperation between farmers and water resp. agricultural authorities, agencies, boards
- Involve farmers in decision making
- Give incentives for and when changing to methods attaining less water consumption

→ Policy option category: Development policies

- Involve environmental, social and economic factors if changing the existent development structure/policy. Agricultural areas form and belong meanwhile to the natural environment as well as to the social structures of the islands
- Define direction of social policies

2.5 Water Quality

A sustainable and efficient water management does not only depend on water quantity, but also and foremost on the quality of the existing waters, even though water quantity has a higher priority for the Mediterranean area. Yet, water of an inappropriate and bad quality limits water resources which are already low and thus water supply. Water quantity and water quality depend on one another. Even though water quality data are rather rare on the islands under consideration in MEDIS, an overall conclusion regarding the sources of pollution for surface waters, including dams and groundwater, will be given in the following. The main water pollution generates from:

- Natural endogenous salts, thus natural salinization due to the hydrogeomorphology,the main natural polluters are: boron for Cyprus, gypsum for Crete, various metals for Corsica and various salt reach terrains for Majorca (triassic evaporates, tertiary saline deposits)
- Sea water intrusion, thus salinization of aquifers due to an overexploitation of the coast near aquifers
- Agricultural methods due to overusing fertilizers and pesticides which often leachate into the aquifers
- Solid waste disposal; and last but not least
- Wastewater treatment plants with regard to nitrates, phosphates and pathogenous microorganisms/germs

→ **Policy option category: Supply management/Water quality/Environmental policy**

- Reduce amount of agrochemicals – fertilizers and pesticides – applied by employing integrated crop management, officially prescribing a recommended dose, documenting the amount of agrochemicals applied and adjusting them to the real crop's needs, by educating informing and training farmers, distributing leaflets with the appropriate information, monitoring the soil, soil water and groundwater near aquifers with regard to salts and pesticides, by carrying out studies on the factual agrochemical applications on the fields, analysing the soil before applying the appropriate amount of agrochemicals in order to reach the best development of the crops and the lowest salt leaching and by penalizing excessive uses. All measures to be taken are in fact the implementation of the EU-WFD.
- Reduce in reasonably restricted quantities the solubility of natural salts, for example keeping the pH (acidity) of the soils within the appropriate range, or setting up strict standards for the elimination of anthropogenic sources of the pollution from natural salts
- Restrict and keep tight control of water abstraction from coast near aquifers to avoid further salinization
- Construct appropriate treatment plants for the island's situation and improve the existing plants
- Monitor regularly the wastewater treatment plants, their effluents

→ **Policy option category: Institutional policy/Capacity building**

- Inform the population with regard to their use of water and wastewater to improve the effluents quality of the treatment plants

2.6 Indicators of Socio-Economic Conditions

2.6.1 Population and Population Growth

The main characteristics of each island concerning the island's size and population data are given in Table 2.1. It should be noted here that Majorca, with the second highest population density is the smallest island and receives the most tourists, up to 4 times more than the other islands, which leads to high domestic water consumption. Furthermore, Majorca also consumes high amounts of water for agriculture. Above all, its low mean water availability worsens the situation and puts pressure on Majorca's water supply. Sicily has the highest population density due to a great "city-density". Sicily has big, crowded cities which are densely populated entailing a high consumption of water, which explains the high record of the indicator "water consumption per unit area," as shown in the indicators for the conditions of the natural environment.

With regard to the population growth UN-projections reveal an increasing trend of about 1% a year of the world's population. Yet, population growth in Europe regresses by 0.2% a year (FAO-website, statistics). Regarding water management on a regional scale the local situation of population growth is therefore of greater importance than generalized statements. In general the population growth in the five Mediterranean countries - Spain, France, Italy, Greece, and Cyprus - shows an increasing tendency. Yet this rise has been different from island to island varying between 3.9–41% over the last 42 years (Table 2.7), with Sicily showing a decrease of 0.4% according to the data of the last three years. The island population growth does not always follow the national population growth. For example, while Spain shows a 4% growth, Majorca has a population growth of 41% (Table 2.7).

On the one hand this indicator is very important for water consumption estimations and scenario plans for the water availability in future and actions to be carried out. On the other hand every indicator should be correlated to others in order to achieve an overall assessment and understanding of the situation, as the following example from Cyprus illustrates.

Although the population on Cyprus has increased, the water consumption for domestic purposes has reduced by 20% since 1990. In addition, the decrease in irrigation accounted to 67%. The reduction in irrigation water during the 1990s was achieved due to climatic droughts which induced drastic falls on annual crops; priority was given to permanent plantations.

While the indicators for the natural environment where considered, it was shown that the water availability and exploitability for Cyprus and Majorca was very low. However, when the population is considered and related to the available water resources, it is evident that all islands lie under the threshold level of 1,700 m^3/person and year which stands for an irregular and local water shortage, with the exception of Corsica and Crete (Fig. 2.20). When the exploitable water resources are considered Crete also falls below the threshold level and faces the aforementioned local water shortages; Majorca, Sicily and Cyprus have to deal with crucial water shortages and with water scarcity. The person related water exploitability of these three islands decline once more reaching values below 500 m^3/person^{a}year which is a main constraint to life.

2.6.2 *Gross Domestic Product – Employment*

An important indicator pertaining to the socio-economic situation and characterizing the economic value of a country and thus its social structure is the Gross Domestic Product (GDP), as well as the employment of each sector contributing to it. The GDP is the total value of all final goods and services produced in a country and is therefore the specific indicator of the country's productivity and the contribution of the various working sectors to the economy. The working sectors are divided into the primary, the secondary, the tertiary and the quaternary sectors. The branches they include are the following: The primary sector is the sector in which

Table 2.7 Population growth in Spain, Majorca, France, Corsica, Italy, Sicily, Greece, Crete and Cyprus (Sources: Spain and Majorca: National Institute of Statistics, INE 2004; Cyprus: Republic of Cyprus, Statistical Service 2003; Greece and Crete: National Statistical Service of Greece; France and Corsica: INSEE 200 SIRENE repertory; Italy and Sicily: ISTAT, all data in: Socio-economic report, Cyprus team, 2004.)

Year	Spain1 1981–2003	Majorca 1981–2003	France (mid. yr.) 1985–2003	Corsica 1985–2003	Italy 1999–2001	Sicily1 1999–2001	Greece 1961–2001	Crete19 1961–2001	Cyprus[a] (Total) 1992–2002	Cyprus[b] 1992–2002
Pop. Grow. over:	5% 22 years	41% 22 years	4.9% 18 years	3.9% 18 years	0.4% 2 years	–0.4% 2 years	30.74% 50 years	24.4% 50 years	13% 10 years	16.3% 10 years
Pop. Grow. per year	0.23%	1.9%	0.27%	0.22%	0.2%	–0.2%	0.6%	0.49%	1.3%	1.6%

[a] Entire island of Cyprus
[b] Governmental controlled area

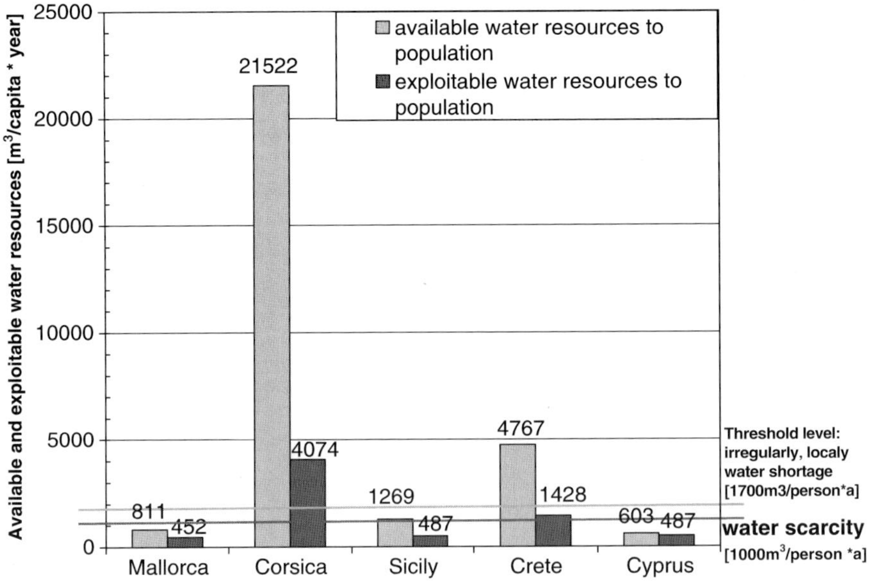

Fig. 2.20 Volumes of available and exploitable water resources per capita and year

raw materials are extracted from the natural environment, including mining, farming (agriculture), fishing and forestry. The secondary sector is manufacturing, i.e., processing of raw materials and generating of energy/power production, building and construction, and all branches of industry. The tertiary sector provides services such as teaching and nursing, transports and communication, enterprise and firms, commerce and banks, tourism and other social services, bakeries, bookstores, etc. The quaternary sector includes research and development. Table 2.6 depicts the GDP of the islands and the respective country they belong to, the average GDP per capita, the average per capita income, the contribution of each working sector to the economy, the personal and household income on the islands as well as the employment situation in each sector.

Even though each of the above data refers to the years 2000, 2001 and 2002 a comparison can be made within reasonable restrains:

Considering the income per person, Corsica and Majorca reach the highest level with €19133 for Corsica and €19 362 for Majorca. In general Crete has the lowest economic values (GDP, income, etc.) among all islands considered within this survey. Regarding the contribution of the various working sectors to the local economy, agriculture is the sector where the economic contribution is the lowest for almost all islands, ranging from 2% to 6% with the exception of Crete whose input reaches up to 13%. The contribution of tourism is higher with 9–10% for all islands with the exception of Majorca reaching 29%. The employment rates in agriculture are comparatively low as well, these of the touristic sector up to double as high. Majorca represents an exception, employing almost half of its working population in the touristic sector.

Table 2.8 Socio-economic profile of the islands under consideration in MEDIS. (From Socio-economic Report, Island team Cyprus, 2004 and the island teams.)

	Majorca[a]/Sp.	Corsica/Fr	Sicily/It	Crete/Gr	Cyprus
GDP of the country [€]	464,250 $^{a}10^{6a}$ (2000)	1,497,081 $^{a}10^{6}$ (2002)	921, 370 $^{a}10^{6}$ (2001)	130, 436 $^{a}10^{6}$ (2001) (market prices)	9, 559 $^{a}10^{6}$ (2000)
On the island [€]	13,792 $^{a}10^{6}$ (2000)	5,052 $^{a}10^{6}$ (2002)	67,494 ^{a} 10^{6}	6,930 $^{a}10^{3}$ (2001)	9,559 $^{a}10^{6}$
Per capita (island) [€]	19,362 (2000)	19,133 (2002)	13,266	11,533.38 (2001)	13,932
Declared income per inhabitant a year [€]		1,400/month	12,000/a	3,844.5 (2001)	13,000
Average household income [€]	14,752 (2003)	11,243/a	1,395/month	11,539.6 (reported)	
Employed/1000inh.	543.8	319.3	260	893	931
Contribution of agriculture to the island GDP	2%	2.4%	6.1%	12%	4%
Employment agriculture	2.2%	4.4%	9.5%	32.9%	8.6%
Contribution of industry (incl. construction) to the island GDP	16%	13.6%	20.5%	11%	21.9%
Employment – industry	24.9%	15.2%	17.7%	10.3%	21.4%
Contribution of services to the island GDP	86%	84%	73.4%	77%	63.2%
Contribution of tourism to the island GDP included in services	29%	9.5%			10.4%
Employment-tourism	40.5%	12.5%		9%	10.9%

[a] Balearic Islands

The ratios of the islands economic value (GDP) in a working sector to the employed population illustrates the efficiency and productivity in the corresponding sector. Regarding Majorca's agricultural sector, its ratio of 1:1.1 indicates that agriculture is pursued effectively and productively and Sicily with a ratio of 1:1.6 operates almost as effectively. In contrast, Corsica with a ratio of 1:1.8, Cyprus with 1:2.2 and Crete with the greatest ratio of 1:2.7 signify a dire need to improve their agricultural management. The appropriate ratios of the industrial sector all vary between 1:0.8 and 1:1.6 (Majorca) and thus imply a high efficiency. Similar records can be found within the service sector. The ratios for the touristic sector (part of the service sector) range from 1:1 for Cyprus and 1:1.4 for Majorca demonstrating an increased efficiency.

Even though agriculture belongs to a working sector whose economic contribution to the islands' economy does not attain the level of the other sectors, and employment in this field not significantly contributing to the overall employment situation – although Crete's labour force employed in agriculture is one third of the employed population – agriculture constructively influences the above mentioned fields. Cyprus expresses this as follows: "The broad agricultural sector, despite the reduction of its contribution to the GDP and total employment, continues to be a fundamental sector of the Cyprus economy, both with respect to their production of essential food items for the population and exports and with respect to the employment of thousands of rural residents and the containment of the depopulation of the villages" (Report socio-economic analysis).

Based on this statement which is in conformity with the majority of the other islands, agriculture, as the main water consumer which does not have a major contribution to other fields such as the economy or employment market and compensate in a way for the water stress it creates, should aim at becoming more efficient and productive in addition to its water-saving measures which were previously mentioned. In this context few more options are worth mentioning:

→ **Policy option category: Institutional policies/capacity building**

- Improvement of the agricultural sectors for more efficiency and productivity which is in accordance with the quality and quantity of water resources
- Ensure fairness and right of existence to all working sectors
- Define direction of development priorities coherent to all sectors and stakeholders

2.6.3 Water Price

Through the Water Framework Directive the European Union provides various tools in order to achieve its goal for reaching and maintaining good water quality. One of these tools is the use of an adequate pricing policy. All five islands under consideration in MEDIS have a different pricing system. There are different models of progressive or proportional water pricing, municipalities implementing their own water rates, a number of consumers paying only when they are connected to the Municipal Distribution Network, or consumers that own a well not paying anything or only for their infrastructure maintenance and operation including sanitation taxes. Furthermore, the water prices themselves vary not only among the islands but also throughout an island depending on use, location and pricing body, catchment and category of consumer and within the same city or catchment, supplier, town or city. The prices vary between €0.015/m^3, the lowest price in Crete up to €2.57/m^3 for Corsica, for domestic use and between €0.017/m^3 the lowest price in Cyprus up to €0.5/m^3 for Corsica for irrigation use. Wastewater is being supplied for free or for up to €0.069/m^3.

If by coincidence pricing the water is decided by all stakeholders, a homogeneous, justifiable and feasible system should be implemented. The price should be differentiated between the different user groups, yet similar within one group and catchment or city. Care should be given to the price of water for agricultural purposes due to the sensitivity of this matter. Even though water is a common good and should thus not be charged, the provided "water services" such as processing, collecting, cleaning, storing, distributing, treating wastewater and sewage as well as the generated environmental costs etc. should be priced sensibly. Two additional effects emerge through water pricing: A high water quality can be ensured and a reduction in water consumption can be obtained.

→ **Policy option category: Institutional policies/Economic policy/Capacity building**

- Implementation of a sensible, justifiable and feasible pricing system differentiated among the user groups (see above)
- Raise public awareness of the value and scarcity of water by carrying out educational campaigns, setting up educational programs and training at all levels of society

2.7 Conclusion and Perspectives

Water management is a challenging task, especially in arid and semi-arid regions where water availability and exploitability depends on a number of factors such as climatic but also socio-economic and cultural conditions. The indicators presented here dealt with the impact of water availability due to climatic changes as well as the challenge of water usage by conflicting water demands among various sectors. The characterization and comparison of the five islands under consideration in MEDIS have led to a variety of policy options which are in accordance with the EU-WFD and ought to be implemented as soon as possible. One position should explicitly be underlined here. All policy options suggested with regard to the agricultural sector as the main water consumer do not aim at abolishing parts of the agricultural sector, but instead wish to improve this sector and enable an integrated consideration of the situation, such as is evident from the socio-economic values.

Agricultural products belong to human life and the workers within the agricultural sector contribute to the labour market and the GDP. Furthermore agricultural surfaces define a part of our landscapes and shape the natural environment. The point of interest is to see how water, an irreplaceable, natural source, can be kept uncontaminated and how a sustainable use of water for all needs can be achieved. With these priorities in mind, the agricultural sector needs to take over the necessary responsibility. In addition, to ensure a sustainable water management, all sectors, including the domestic one, require a responsible awareness in their usage of water. Moreover, tourism's large contribution to the economy is of great importance yet, nevertheless, the appropriate measures should be carried out here as well in view of a greater responsibility towards the environment. In doing so, all sectors pay tribute to an integrated protection of the water resources.

References

Agricultural Report, Island team Crete (ed.) (2003), '*Evaluation of major cultivated crops/cultivars on catchment and island scale*', Deliverable 7. European Project MEDIS EVK1-2001-00092

Bolle, H.-J. (2003). Climate, climate variability and impacts in the Mediterranean area: An overview in Mediterranean Climate-Variability and Trends; edited by Bolle, H.-J.: Volume 1 in Regional Climate Studies, Bolle, H.-J., M. Menenti and I. Rasool (eds); Springer-Verlag, Berlin, Heidelberg, New York, 5–86

Brouwer, C. and Heibloem, M., FAO (eds.) (1986). Irrigation water management: Irrigation water needs, FAO, Rome, Italy

Chartzoulakis, K.S, Paranychianakis, N.V., Angelakis, A.N. (2001). Water resources management in the island of Crete, Greece, with emphasis on the agricultural use. Water Policy, 3: 193–205

CyStatWebsite, http://www.mof.gov.cy/cystst/statistics.nsf/index en/index en?OpenDocument and http://www.mof.gov.cy/cystst/statistics.nsf/populationcondition en/populationcondition en?Opendocument

Donta, A., Arrighi, M.-E. (ed.) (2003a). Island Report Corsica, in Lange, M. (ed.) First Annual Report of MEDIS, European Project MEDIS EVK1-2001-00092

Donta, A., Island team Crete (ed.) (2003b). '*Island Report Crete*', in Lange, M. (ed.) First Annual Report of MEDIS, European Project MEDIS EVK1-2001-00092

Donta, A., Constantinou, G., Katsikides, S. (ed.) (2003c). Island Report Cyprus, in Lange, M. (ed.) First Annual Report of MEDIS, European Project MEDIS EVK1-2001-00092

Donta, A., Candela, L., von Igel, W., Gallimont, A., Bejarano, C., (ed.) (2003d). Island Report Majorca, in Lange, M. (ed.) First Annual Report of MEDIS, European Project MEDIS EVK1-2001-00092

Donta, A., Aronica, G., (ed.) (2003e). Island Report Sicily, in Lange, M. (ed.) First Annual Report of MEDIS, European Project MEDIS EVK1-2001-00092

EEA (1999). Environmental indicators: Typology and overview, Technical Report No. 25. European Environmental Agency (EEA), Copenhagen

EEA (2003a). EEA core set of indicators, revised version April 2003, Technical Report European Environmental Agency (EEA), Copenhagen

EEA (2003b). Europe's water: An indicator-based assessment, Topic report 1/2003. European Environmental Agency (EEA), Copenhagen

FAO, Http://www.fao.org

FAO & WDD (2002). Re-Assessment of the Water Resources and Demand of the Island of Cyprus, Cyprus, CD-Rom

OECD (1999). Compendium of environmental statistics

Rambaud, Meteo France (2003). '*Les précipitations en Corse*' ("rainfall in Corsica"). In: Convention on Water: Synthesis of the works, Hydraulic Agency of Corsica (OEHC); Territorial Collectivity of Corsica, CD-ROM

Report Institutional Analysis, Island team Cyprus (ed.) (2004). Institutional analysis of water management practices on a catchment scale and on island scale Deliverable 12, European Project MEDIS EVK1-2001-00092

Report Socio-economic analysis, Island team Cyprus (ed.) (2004). Detailed empirical analysis of the present socio-economic structure under the aspect of water usage Deliverable 5, European Project MEDIS EVK1-2001-00092, 2004

Veraart, J.A., De Groot, R.S., Perelló, G., Riddiford, N.J., Roijackers, (2004). Selection of (bio) indicators to assess effects of freshwater use in wetlands: a case study of s'Albufera de Mallorca, Spain, *Regarding Environmental Change*, 4: 107–117

Walmsley, J. (2002). *Framework for measuring sustainable development in catchment systems*, Environmental Management, 29(2)

Water Framework Directive (WFD) 2000/60/EC of the European Parliament and of the Council of 23 October 2000 establishing a framework for Community action in the field of water policy

Xoplaki, E. (2002). Climate variabilità over the Mediterranean, PhD, University of Bern and University of Thessaloniki, Bern Switzerland and Thessaloniki, Greece, 213

Xoplaki, E., Gonzalez-Rouco, J., Luterbacher, J., and Wanner, H. (2004). Wet season Mediterranean precipitation variability: influence of large-scale dynamics and trends, Climate Dynamics, 23: 63–78

Chapter 3
The Range of Existing Circumstances in the WaterStrategyMan Case Studies

Bernard Barraqué, Christos Karavitis, and Pipina Katsiardi

3.1 Introduction

Freshwater is no longer taken for granted as a plentiful, always available, resource. More and more people in an increasing number of countries are experiencing scarcity situations, particularly revealed by droughts. EU Member States are not an exception. Today, many European countries are subject to waves of water deficit that affect their population and the ecosystems they depend on. Recent events have further demonstrated how socio-economic factors, driving the demand for water, have made even the rainiest parts of Europe vulnerable to drought.

Indeed what droughts increasingly tend to reveal is the overexploitation of water resources in some European countries. Increased abstractions, especially for agriculture, create the risk of water deficit and, consequently, environmental hazards. The problem of water deficit resulting from resource overexploitation is further exacerbated by global warming that is likely to make precipitation patterns more variable, changing water availability conditions in Europe on quantitative, temporal and/or seasonal basis. Alternative approaches are therefore needed to meet the water requirements for development activities while leaving enough resources for the sustainability of the aquatic ecosystem. These new approaches are being driven by a growing awareness of the values brought about by an adequate availability of water.

A new strong element that should be taken into account in the decision-making process is the widely adopted Water Framework Directive. The WFD 2000/60/EC addresses primarily the issue of water quality and environmental protection through policy instruments that put emphasis on:

- Developing administration at the river basin district level
- Decentralising decision making, (principle of subsidiarity)
- Monitoring quality and progress achieved through a transparent reporting system
- Planning and integrated water resources management
- Developing public participation and citizen empowerment
- Getting closer to cost recovery for water services and rationalization of water use
- Appropriate allocation of costs to users and between users

P. Koundouri (ed.), *Coping with Water Deficiency,* 45–112.

Quantitative issues are mainly tackled in Article 1 of the WFD by promoting a sustainable use of water resources. They are also addressed in Article 17, requiring a daughter Directive for the management of groundwater. Although they are explicitly mentioned for groundwater, they remain an indirect concern with respect to surface water. Several Member States and the Commission consider the need to develop a more explicit strategy concerning water scarcity and drought, like they did earlier for floods.

The WaterStrategyMan project aims at contributing to solving problems of water shortage in arid and semi-arid regions of Southern Europe, where conditions and issues differ from those in the Northern Europe due to precipitation variability and the importance and conflicts between tourism and irrigation. The project seeks to develop and evaluate strategies and guidelines towards integrated water resources management in the Southern European Regions. The adopted research approach is based on the successive generalization resulting from systematic analysis of specific conditions. On the basis of a generalized conceptualization of scarcity, the analysis of specific regions aimed at providing knowledge that was used to develop Regional Models, which will be further analyzed through the Case Studies. The Case Studies, through the use of a developed Decision Support System, are expected to subsequently yield results that can be used for a broad spectrum of similar regions. For instance, in France, the demand for irrigation water increases, while in several regions, which are not necessarily in the Mediterranean part, this demand cannot even be met in moderate drought conditions.

Additionally, in most European countries, solutions to shortage problems, based on civil engineering, water storage and long distance transfers, are increasingly questioned because of their ecological impact as well as their cost. This leads, of course, to a demand-side solutions; however, these will depend both on natural and man-made contexts of shortages.

There are variations over the definition of aridity and the boundaries of arid and semi-arid regions. The terms "aridity" and "water deficiency" are not interchangeable; yet, they apply to a large proportion of the Mediterranean countries. In order to provide an overview of the conditions in southern European regions, six participating countries were examined; regionalization was a key step for the selection of a representative sample of regions. It was important to select a suitable range of regions in terms of water deficiency, to ensure that the outcome of the analysis can eventually apply to the widest possible range of water deficient areas, in order to emphasize the regional character of water shortages, and to study those particular areas through characteristic case studies. The regions analysed were selected based on the following criteria, or their combinations:

- Existence of natural aridity in the areas
- Existence of water shortages on a permanent or seasonal basis due to natural or man-made reasons, or the recurrence of drought and/or flood cycles
- Insufficient efforts of water resources management in the areas
- Lack of proper administrative or institutional framework for an effective water resources management
- Socio-economic conditions of the areas, that affect the management of water resources.

To that end, 15 regions were selected in the six participating countries. Water stress conditions and the spectrum of circumstances were analysed in a country level and within the specific regions. The 15 selected regions were the following:

- Greece, Attica, Thessaly and the Cyclades Islands
- Italy, Belice basin and Emilia-Romagna
- Spain, Doñana and the Canary islands
- Portugal, Algarve, Sado and Guadiana
- Cyprus, Akrotiri, Germasogeia and Kokkinochoria
- Israel, Tel Aviv and Arava regions.

The main observation was that aridity should be understood in a wider context encompassing both natural and man-made processes. Various concepts have been used to exemplify a prevailing confusion among such terms which signify dry environments or water deficiencies. The terms vary all the way from the extreme of desert to aridity, to drought and to temporary water shortage. There are four different terms that are important to the initial separation between physical and social conditions with regard to what one can summarily label water deficiencies:

- **Aridity**, signifying a permanent natural condition and a stable climatic feature of a given region
- **Drought**, referring to a temporary feature of the climate or to regular or unpredictable climatic changes
- **Water shortage**, a term that can be understood mostly as a man-made phenomenon reflecting the concern of temporary and small area water deficiencies
- **Desertification**, as a process of alteration of the ecological regime often associated with aridity and/or drought but principally brought about by human activities which change significantly the surrounding ecosystem.

The WaterStrategyMan regions presented in this chapter are described through several elements: climate, water availability and quality issues, demography, relative

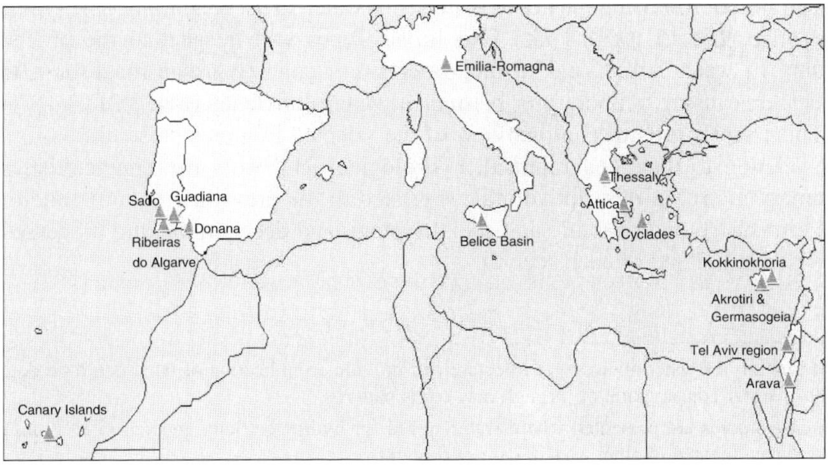

Fig. 3.1 The 15 selected regions in the six case study countries

importance of various types of stakeholders (domestic water supply, additional tourist demand, agriculture,[1] hydroelectricity,[2] industry, the ecosystem itself), pricing and policy issues. They provide as many case studies for discussing the move towards demand management. There is one broad and striking opposition that can be noted from the start, between the mainland areas and the islands: in the case of the latter, there is no tradition in solving relative scarcity problems through importation of external resources, for lack of an appropriate technology (such as flexible pipes lying on sea bottom); besides, there are usually few sites available for building large (interannual) reservoirs, and not enough rainfall. This is notably why Cyprus and Greek islands took a certain lead in seawater desalination technology. Conversely, there are regional or national bulk water conveyors in Israel, Attica and Thessaly, and southern Portugal. This kind of water transfers can also be met in other areas like Puglia in southern Italy, several regions in Spain, and in Provence and Languedoc, the two Mediterranean regions of France. This is why both islands and mainland regions were chosen for case-studies. However, as it was mentioned above, even external supply-side solutions are increasingly accompanied by new pricing schemes. These are supposed to reflect part of the investment cost, which furthermore creates a demand for control. Innovative responses to water deficiency problems, technological and social, as well as legal mechanisms for carrying out management schemes tend to fall under the following four major categories:

- **Strong incentives for efficient or new uses**, including economic benefits, redefinition of the doctrine of beneficial use, etc.
- **Institutional changes**, such as new organizational arrangements, creation of new water agencies, development of bargaining between water uses, etc.
- **Regulatory counter-incentives**, such as stricter enforcement and taxation/pricing policies
- **Changes in water lifestyles** and cultural practices, and changing crops.

On the basis of the spectrum of water management circumstances and the DPSIR causal interrelationships in the 15 regions, a typology for water stress conditions was proposed, following the above analysis of water stress conditions. The typology categorizes the 15 regions into four broad Types with respect to the processes leading to water stress, and to the water stress context – man-made or natural processes, causing temporary or permanent water deficiency (Fig. 3.2).

This chapter provides an overview of the selected Regions, presenting information relating to their geographical, hydrological and water use characteristics; a summary matrix of descriptive indices related to the prevailing natural conditions and infrastructure, economic and social system, and decision making processes, is provided at the end of each section.

---

[1] The quantity of water allocated per hectare can provide an indication of what could be done in terms of water conservation or through new crops patterns.

[2] No case studies are presented where water stored for hydroelectricity generation in autumn or winter is reallocated to other uses in case of drought.

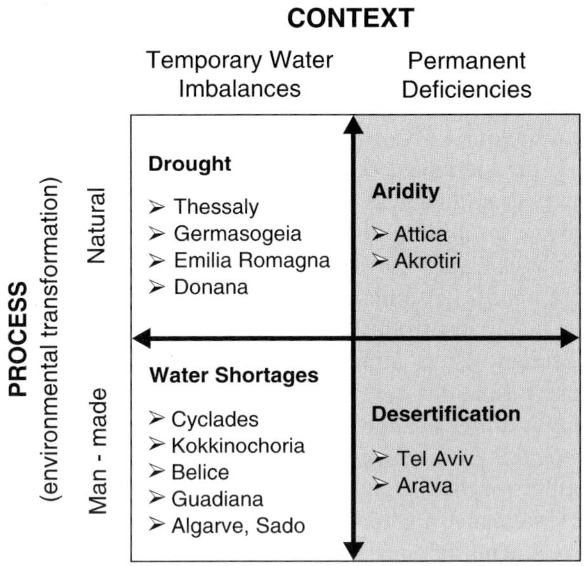

Fig. 3.2 Typology of the selected regions

The information available in this chapter has been collected and provided by the WaterStrategyMan Case Study Project Partners:

- The National Technical University of Athens, Greece (Professor D. Assimacopoulos)
- ProGEA S.r.l., Italy (Professor E. Todini)
- The Hebrew University of Jerusalem, Israel (Professor E. Feinerman)
- The Water Development Department and Aeoliki Ltd., Cyprus (Mr. I. Iacovides and Mr. N. Nicodemou, and Dr. I. Glekas respectively)
- INSULA, Spain (Mr. C. Marin)
- The University of Porto, Portugal (Professor R. Maia).

3.2 Attica Region, Greece

The Water Region of Attica covers an area of 3,207 km^2. The region has several mountains (Parnitha, Kitheronas, Penteli, Imitos, Egaleo, Pateras) and plains on the coastal zone. The average elevation is 115 m. It also includes a few islands – with Egina, Salamis and the uninhabited Makronissos being the biggest among them – plus a small part of Sterea Ellada (Viotia) and Peloponnesus.

Attica borders with Sterea Ellada to the north, Peloponnesus to the south. Its eastern shoreline is on the south of Evoikos Gulf and the Aegean Sea and its western shoreline is on the Saronikos Gulf.

The two main rivers are Illissos and Kifissos, which, in the urbanized areas, have been transformed into covered stormwater conduits and drain into Saronikos Gulf.

Their drainage basin covers 320 km². There are two small natural lakes, Vouliagmene and Koumoundourou. None of the drainage basins in Attica is larger than 600 km². The reservoir of Marathon, which is used for the water supply of the metropolitan Athens area, is located in the drainage basin of the Charadros River (185 km²). The capacity of the reservoir is 41 hm³.

The climate is Mediterranean continental. The average temperature is 16–18 °C. The average annual rain height is 400 mm, ranging from less than 400 mm in the south coastal areas, to 600 mm at the mainland and 1,000 mm on the mountains. Rain frequency varies from 50 to 100 days per year. The precipitation is 1,698 hm³/year and the total runoff 259 hm³/year. Snowfalls are very rare at the coastal areas but occur on the mountains from October to April.

Permeable geological formations cover a significant amount of the total area. Karstic limestone formations cover the east and west part of the region. The total water availability is about 449 hm³. This amount consists of 259 hm³ surface water and 190 hm³ potential groundwater. The groundwater can be found in the karstic and alluvial aquifers of the region.

There is no systematic monitoring of surface water quality in Attica, but since rivers are the recipients of unprocessed wastewater, they are generally in poor condition and their exploitation is unattainable. The treatment plant of Psytalleia

Table 3.1 Surface of the drainage basins in Attica

Drainage basin	Surface (km²)
Attica's Kiffisos river – Illissos river	320
Sarantapotamos river	310
Charadros river	185

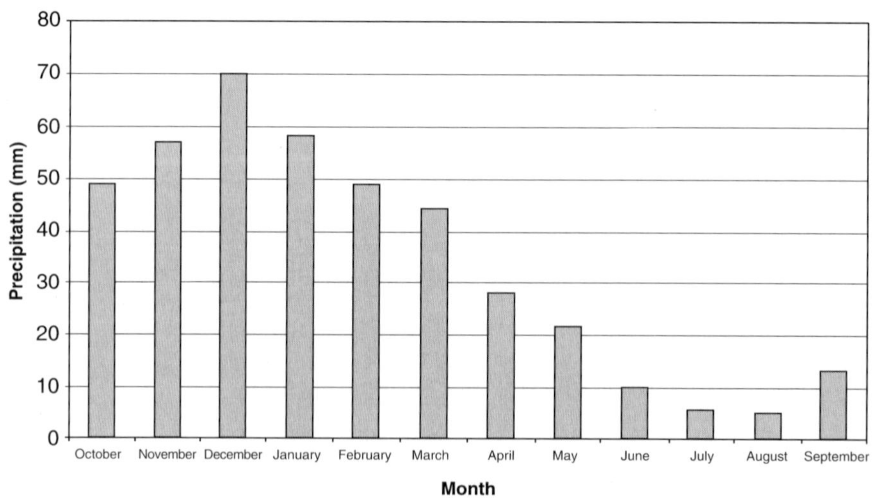

Fig. 3.3 Mean monthly precipitation for the period 1955–1995 in Attica

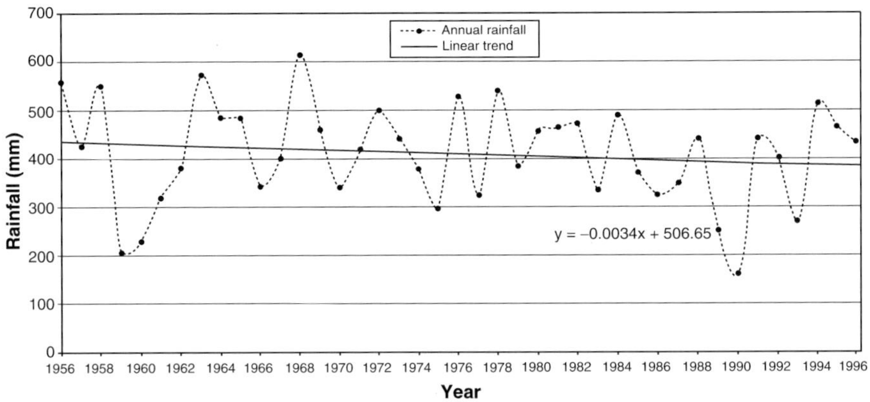

Fig. 3.4 Annual rainfall in Attica

Table 3.2 Hydrological entities of Attica

Hydrological entities	Surface (km^2)	Total runoff (hm^3/yr)	Potential groundwater (hm^3/yr)
Karstic entities in limestone formations			
South Parnitha – East Pateras – Egaleo	510	157	120
Kitheronas	260	75	50–70
Gerania	250	42	20
Penteli	250	55	30
Ymitos	110	15	30
Northeastern Parnitha	300	95	60
Total		439	250–2704
Alluvial aquifers			
Athens	440	30	5
Mesogeia	820	50	15
Megara	260	15	3
Loutraki	320	20	4
Total		115	27

treats 80–90% of the produced domestic wastewater. Many of the industries use the sewage system as well, but some others, through illegal connections, throw their waste directly into rivers and torrents or into the sea, which results in the degradation of the quality of surface and ground water.

According to ground water results nitrates exceed the critical load for drinking water. In certain areas with significant industrial activity, high concentrations of heavy metals occur.

Table 3.3 Attica matrix of circumstances

Natural conditions and infrastructure		Climate type	Mediterranean
	Regional context	Aridity Index	0.31– Semi-arid
		Permanent population	3,761,810
	Water availability	Total water resources / Availability (hm3)	449
		Trans-boundary water	388
		Quality of surface water	–
	Water quality	Quality of groundwater	Poor
		Quality of coastal water	Fair – Poor
	Water supply	Percentage of supply coming from:	
		Groundwater	17%
		Surface water	–
		Desalination, Recycling	–
		Importing	83%
		Network coverage:	
		Domestic	100%
		Irrigation	–
		Sewerage	90%
Economic and social system	**Water use**	Water consumption by category:	
		Domestic	71%
		Tourism	–
		Irrigation	25%
		Industrial and energy production	4%
		Population to resources index	4494
	Water demand	Water Demand trends	Decreasing
		Consumption index	69%
		Exploitation index	49%
	Pricing system	Average household budget for domestic water (pa)	€117
		Average household budget for agricultural water	
		Average household income	€ 20,639
		Cost recovery	Good
		Price elasticity	Fair
	Social capacity building	Public participation in decisions	Poor
		Public education on water conservation issues	Fair
Decision making process	**Water resources management**	Water ownership	State
		Decision making level (municipal, regional, national) regarding:	
		Water supply for each sector	National
		Water resources allocation for each sector	National
	Water policy	Local economy basis	Tertiary sector
		Development priorities	Urban growth

The main pollutant loads produced in Attica for the year 1996 were estimated as:

- BOD5: 120,000 ton/year
- TSS: 420,000 ton/year
- Total nitrogen: 20,000 ton/year
- Total phosphorus: 8,000 ton/year

The prefecture of Attica has 3,761,810 inhabitants (2001). It includes the capital district, small parts of Sterea Ellada and Peloponnesus Region. The main economic activities are commerce, industry, agriculture and tourism. The region produces 36% of the GNP, while the per capita product is €12,560, and the mean declared income per inhabitant was €6930 in 2000. The unemployment rate in the region is 10.4%. The total annual water demand is 408 hm^3, consisting of 289 hm^3 for domestic use, 101.5 hm^3 for agricultural use and 17.5 hm^3 for industrial use. The total water consumption in the area represents 5.3% of the total water consumption in Greece. The consumption index is estimated to be equal to 69% and the population to water resources index is equal to 4,494. The exploitation index is 49%.

In order to satisfy the demand, a significant amount of water is imported from the Hydrological Department of West Sterea Ellada (Rivers Mornos and Evinos), and from the Hydrological Department of East Sterea Ellada (Lakes Iliki and Paralimni).

A part of the industrial demand is covered through desalination plants.

The supply of water and wastewater services is effected by EYDAP S.A., the biggest water service provider in Greece. Hence, the cost recovery for water services in this region is good, the pricing of water is done on the basis of the services provided and is not subject to political pressures. Attica is the only part of Greece where demand management through pricing control has been effective, in the drought periods in the 1990s. As a consequence, public education on water conservation issues is on average better than in most other regions of the country, although there is little to no public participation on water-related decisions. Decision making regarding water issues in Attica is effected on a national level, as the region is under the direct control of the Ministry of the Environment.

3.3 Thessaly Region, Greece

The fertile plain of Thessaly water region covers an area of 13,377 km^2 that occupies the central section of mainland Greece. It is surrounded by high mountain ranges with altitudes more than 2,000 m (Pindus, Olympus, Pelion, Othrys, Ossa and Agrapha), encircling a low plain. The River Pinios, coming down from the western slopes of Pindus, cuts Thessaly in two, passes through the valley of Tempi and meets the sea. Thessaly borders with Macedonia to the north, Sterea Ellada to the south, Epirus to the west. Its eastern shoreline is on the Aegean. It has the highest percentage of flat land in Greece and the mean elevation of the area is 285 m.

Among the mountains flows the Pinios River which drains into the Aegean, after passing through the Thessalic Tempi. The drainage basin of Pinios River is 9,500 km^2 and the main tributaries are the rivers Titarisios, Enipeas, Kalentzis, Litheos and Asmaki. Thessaly District consists also of two more hydrologic basins: the drainage basin of Lake Karla (1,050 km^2), rising at the eastern side of the District and Lake Plastira at the western side. Lake Plastira is a part of the watershed area of Achelloos River which belongs to the West Sterea Ellada Water Region.

The climate is Mediterranean continental. Winters are cold, summers are hot, with a large temperature difference between the two seasons. The average temperature is 16–17 °C. The average annual rain height is 700 mm, ranging from 400 to 600 mm at the central plains, to 600–1,000 mm on the eastern part and over 1,200 mm on the mountains. Rain frequency varies from 100 to 130 days per year. The precipitation is 10,426 hm^3/year and the mean annual relative humidity is 67–72%. Snowfalls are very frequent on the mountains.

Table 3.4 Surface of the drainage basins in Thessaly

Drainage basin	Surface (km^2)
Pinios River	9,500
Lake Karla	1,050
Other Basins	2,812
Total	13,362

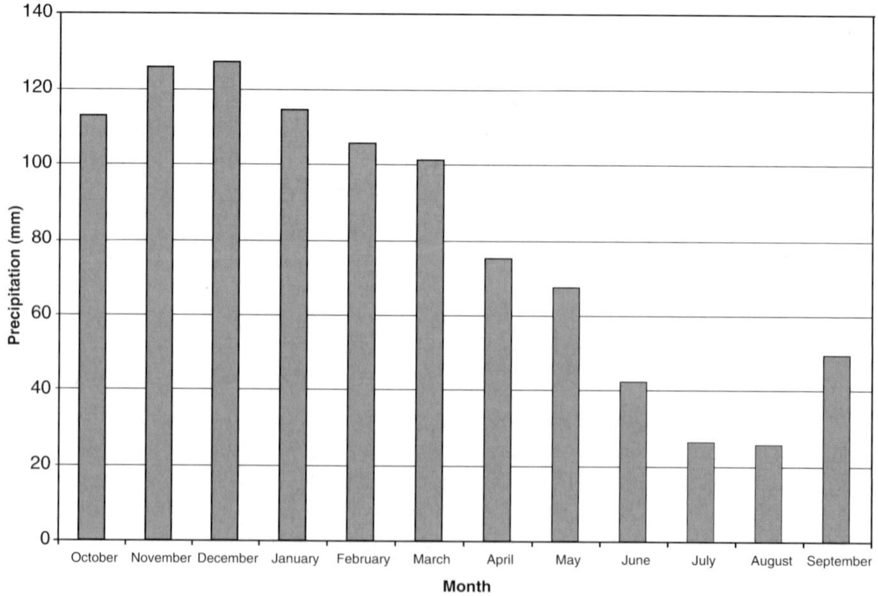

Fig. 3.5 Mean monthly precipitation for the period 1955–1995 in Thessaly

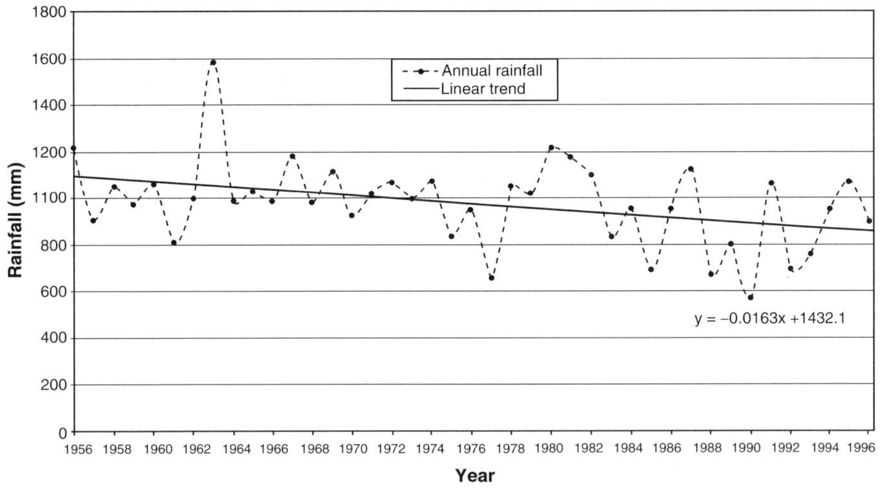

Fig. 3.6 Annual rainfall in Thessaly

Impermeable geological structures cover 39.4% of the total area; karstic aquifers cover 16.2% and permeable structures which occur mainly on the plain cover 44.4%. The total water availability is about 3,094 hm^3. This amount consists of 2,558 hm^3 surface water and 506 hm^3 groundwater. The groundwater, which can be found in the karstic and alluvial aquifers of the region and the entire plain, consisting mainly of Neogene sediments, is fed from Pinios River and the tributaries, as well as from direct rainfall infiltration.

According to monitoring results, surface water quality in Thessaly is generally in a good condition. The nitrite concentrations in a few sampling points exceed the limit values for drinking water, due to the cultivation carried out to serve agriculture in the specific areas of the drainage basin. In few of the sampling points, results for pesticides show elevated levels. Although urban waste loads in the water are significant, urban waste water treatment plants in the major cities of Thessaly ensure that the water quality remains good. The treatment plants constructed in all the major cities of the area are efficient enough and 45% of the population of the area (80% of the urban population) were connected to the public sewerage network in 1998.

Results for groundwater show that in many cases nitrates and in some cases ammonia exceed the critical load for drinking water. Because of this, the Thessaly plain is designated as a vulnerable zone (Joint Ministerial Decision 19652/1906/99), in order to take the appropriate measures for the protection of the area. The elevated nitrate and ammonia levels are attributed to agriculture and animal husbandry practices.

The main anthropogenic pressures observed in Thessaly are caused by loads originating from agricultural and animal breeding activities (significant non-point source loads) and from the urban wastewater. Cultivated areas are spread all over the plain areas and the land-application of all nitrogen-containing fertilizers enriches the watercourses causing significant pollution trends.

Pollution trends caused by industrial activities are not significant because of the limited industrial production. Pollution loads from industry are most abundant in Larissa and Volos where most industrial units are concentrated, and are particularly visible in the coastal waters of the region.

The main pollutant loads produced in Thessaly in 1996 were:

- BOD5: 51,740 ton/year
- TSS: 66,670 ton/year
- Total nitrogen: 37,920 ton/year
- Total phosphorus: 3,750 ton/year

Thessaly has 753,848 inhabitants. The biggest cities in the area are Larissa and Volos (total population of both cities 300,000). The main economic activities are agriculture, industry and tourism. The region produces 6.3% of the GNP of Greece, while the per capita product is €10,950, and the mean declared income per inhabitant was €3,550 in 2000. The unemployment rate in the region is 12.2%. Total annual water consumption is 1,171 hm^3, consisting of 65 hm^3 for domestic use, 1.060 hm^3 for agricultural use and 46 hm^3 for industrial use. The total water consumption in the area represents 18.5% of the total water consumption in Greece. The consumption index is estimated to be equal to 38% and the population to water resources index is equal to 204. The exploitation index is 31%.

Irrigated agricultural land occupies 1,894 km^2. Water shortage problems occur during the irrigation period, while in the winter large areas are flooded. The watercourse is also significantly used in animal breeding and aquaculture.

The coastal zone in the area is a favourite destination for many tourists during the summer, and therefore water supply requirements increase during the summer tourist period. The annual water demands for domestic use and for tourism are about 53.7 hm^3, and the areas with the higher water supply requirements are the municipalities of Larissa and Volos.

Lake Plastira, with a storage capacity of 400 hm^3, is regulated for hydropower production. The installed hydropower capacity is 141 MW and the power plant produces a total of 250 GWh. Industrial activities are limited in the cities of Volos and Larissa, and they are related mainly to food processing, textile works and iron and steel production. As mentioned above, water demands for industry are not significant. Thessaly has a dense network of motorways and a harbour in Volos that serves the entire area.

Water supply in the region is not regulated by a single authority. Each one of the larger cities has its own water and wastewater service provider, but on a local level, those services are mostly provided through the municipalities. Thus, the pricing of water is a subject of political pressures. Public education for water conservation is limited, and cost recovery, with the exception of large cities, is poor.

Table 3.5 Thessaly matrix of circumstances

Natural conditions and infrastructure	**Regional context**	Climate type	Mediterranean continental
		Aridity Index	AI > = 0.65 0.2 < = AI < 0.5
		Permanent population	753,848
	Water availability	Total water resources / availability (hm3)	3,094
		Trans-boundary water (hm3)	
	Water quality	Quality of surface water	Good
		Quality of groundwater	Average
		Quality of coastal water	Poor
	Water supply	Percentage of supply coming from:	
		Groundwater	15.7%
		Surface water	68.3%
		Desalination, recycling	–
		Importing	16.0%
		Network coverage:	
		Domestic	
		Irrigation	45%
		Sewerage	
Economic and social system	**Water use**	Water consumption by category:	
		Domestic	3.3%
		Tourism	95.8%
		Irrigation	0.9%
		Industrial and energy production	
		Population to resources index	204
	Water demand	Water Demand trends	Stable
		Consumption index	38%
		Exploitation index	31%
	Pricing system	Average household budget for domestic water (pa)	€ 149
		Average household budget for agricultural water	
		Average household income	€ 10,582
		Cost recovery	Poor
		Price elasticity	Poor
Decision Making Process	**Social capacity building**	Public participation in decisions	Poor
		Public education on water conservation issues	Average
	Water resources management	Water ownership	State
		Decision-making level (municipal, regional, national) regarding:	
		Water supply for each sector	Regional
		Water resources allocation for each sector	National
	Water policy	Local economy basis	Primary sector
		Development priorities	Agriculture

3.4 Cyclades Islands, Greece

The Water Region of Cyclades covers an area of 2,553 km^2. The region consists of 24 inhabited islands, and is characterized by the fragmentation in several smaller units with different climatic, hydrological and geomorphological parameters. The islands are semi-mountainous with plains and the mean elevation level is 160 m. There are no important rivers due to the small size of the islands, except for some torrential ones. Surface water is very limited on this area.

The climate in general is temperate Mediterranean but varies on each island according to the geographical position, the size and the distance from the mainland. The average temperature is 16.5–19.5 °C. The average annual rainfall is 379 mm for the central and southern islands (Naxos meteorological station) and 349 for the northern islands (Athens meteorological station). The precipitation is 902 hm^3/year. The total estimated runoff is 156 hm^3/year. The amount of evapotranspiration is estimated to be equal to 667 hm^3.

The geological formations that appear on the islands vary significantly, with metamorphic rocks covering a large part of the complex. Limestone formations are very limited, whereas volcanic rocks appear on the islands of Thira, Milos and Kimolos. The total water availability is about 212 hm^3. This amount consists of 156 hm^3 surface water and 55 hm^3 potential groundwater, mostly found in the karstic and grainy aquifers of the District. Due to the small size of the islands, springs are accordingly small and salinization problems are often.

There are no monitoring results for surface water quality in Cyclades. The most important sources of pollution are agriculture, animal husbandry and domestic wastewater. Data on groundwater quality are scarce, while coastal aquifers are often subject to saline intrusion due to their overexploitation.

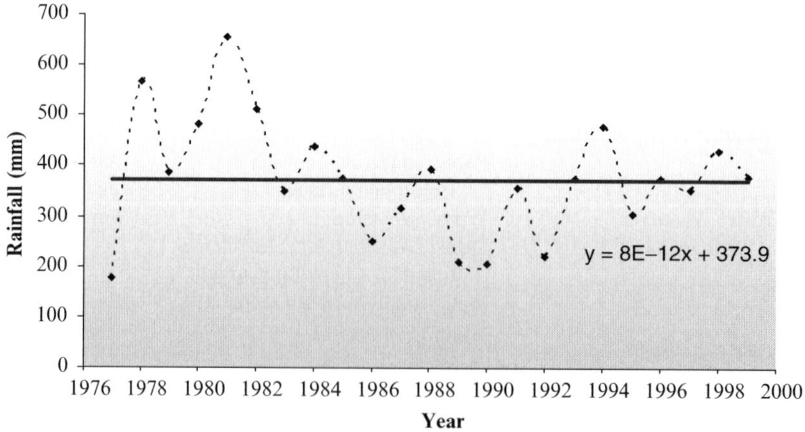

Fig. 3.7 Annual rainfall in Naxos meteorological station

The main pollution loads produced in Aegean Islands in 1996 were:

- BOD5: 8,000 ton/year
- TSS: 9,500 ton/year
- Total nitrogen: 3,400 ton/year
- Total phosphorus: 500 ton/year

The Cyclades have 112,615 inhabitants. The main economic activities are tourism and agriculture. The total annual water demand is 30.95 hm^3, consisting of 7.15 hm^3 for domestic use, 21.5 hm^3 for agricultural use and 2.3 hm^3 for animal husbandry. The total water consumption in the area represents 0.4% of the total water consumption in Greece. The region produces 1% of the GNP, while the per capita product is €12,330 and the mean declared income per inhabitant was €4,600 in 2000. The unemployment rate in the region is 12%. The consumption index is estimated to be equal to 50% and the population to water resources index is equal to 531. The exploitation index is 15%.

In some islands, desalination plants are used to cover water demand and in others water is transferred with tankers from other regions. Several small dams and water tanks have already been constructed and are mainly used for irrigation purposes, while others have been planned and approved to be constructed in the near future.

Water supply in the region is not regulated by a single authority. The larger islands each have their own water and wastewater service providers, but there are a great number of independent local services provided through the municipalities.

Table 3.6 Hydrological data for the Cyclades islands

Island	Surface (km^2)	Precipitation (hm^3/yr)	Evapotraspiration (hm^3/yr)	Total runoff (hm^3/yr)	Ground water (hm^3/yr)
Amorgos	121	45.9	33.89	3.81	8.10
Anafi	38	14.4	10.66	3.64	0.10
Andros	380	144.0	104.75	25.05	2.80
Antiparos	35	13.3	9.77	2.43	1.00
Folegandros	32	12.1	8.88	1.62	1.50
Ios	108	40.9	30.27	10.39	0.24
Kea	131	49.6	36.10	9.50	0.10
Kimolos	36	13.6	10.06	2.24	1.30
Kythnos	99	37.5	27.25	7.15	0.10
Milos	151	57.2	42.33	12.47	2.40
Mykonos	85	32.2	23.83	8.12	0.25
Naxos	428	162.2	120.03	21.72	20.45
Paros	195	73.9	57.76	8.34	7.90
Serifos	73	27.7	20.50	7.10	0.10
Sifnos	73	27.7	20.50	3.00	4.20
Sikinos	41	15.5	11.47	3.23	0.80
Syros	84	31.8	23.53	7.47	0.80
Thira	76	28.8	21.31	5.69	1.80
Tinos	194	73.5	53.48	13.12	1.10

Table 3.7 Cyclades islands matrix of circumstances

Natural conditions and infrastructure	**Regional context**	Climate type	Mediterranean temperate
		Aridity Index	0.3- Semi-arid
		Permanent population	112,615
	Water availability	Total water resources/ availability (hm^3)	212
		Trans-boundary water	Yes
	Water quality	Quality of surface water	–
		Quality of groundwater	Poor
		Quality of coastal water	Good
	Water Supply	Percentage of supply coming from:	
		Groundwater	22.5%
		Surface water	77.5%
		Desalination, recycling	Yes
		Importing	Yes
		Network coverage:	
		Domestic	
		Irrigation	
		Sewerage	
Economic and social system	**Water use**	Water consumption by category:	
		Domestic	9%
		Tourism	14%
		Irrigation	77%
		Industrial and energy production	
		Population to resources index	531
	Water demand	Water demand trends	Increasing
		Consumption index	50%
		Exploitation index	15%
	Pricing system	Average household budget for domestic water (pa)	€ 231
		Average household budget for agricultural water	
		Average household income	€ 13,730
		Cost recovery	Poor
		Price elasticity	Poor
Decision-Making Process	**Social capacity building**	Public participation in decisions	Poor
		Public education on water conservation issues	Average
	Water resources management	Water ownership	State
		Decision making level (municipal, regional, national) regarding:	
		Water supply for each sector	Municipal
		Water resources allocation for each sector	National
	Water policy	Local economy basis	Tertiary sector
		Development priorities	Tourism

Thus, the pricing of water is subject to political pressures. Public education for water conservation is limited and cost recovery is overall poor.

3.5 Emilia-Romagna Region, Italy

Emilia-Romagna is one of the largest Italian regions (the sixth), linking the north with the centre of the country. Stretching as far as the Adriatic Sea to the east, Emilia-Romagna borders with Veneto to the north-east, with Lombardia to the north and north-west, with Piemonte and Liguria to the west, with Tuscany to the south and with the Marche and the Republic of San Marino to the south-east. There is close coincidence of administrative and physical boundaries, delineated by easily distinguishable natural features: the Po River to the north, the Apennine ridge separating the Po Valley slopes from the Tuscany-Marche to the south, and the Adriatic coast to the east.

The region is divided into two main areas: Romagna occupies the south-east with the provinces of Forlì-Cesena and Rimini while Emilia consists of the administrative provinces at the west and middle. Almost the half of the territory is part of the Padana plain, which the Via Emilia separates from the Apennine watershed. A series of nearly parallel ridges thrusts outwards towards the plain, progressively decreasing in height, sharply separated from the transverse river valleys. Beyond the extreme outlying hills, lie the undulations of the stony upper plain, formed by the fusion of fluvial detritus, beyond which extends the wide fertile alluvial plain. Of the great swamps, which at one time characterized the lower Emilia and Romagna plain before systematic regulation of the waterways, remain only the Valleys of Comacchio and the stretches of water belonging to the Po Delta. Except for the Po River, which flows along the northern boundary of the region, all water courses flow from the Apennine watershed, cutting parallel down hill before reaching the plain and flowing into the Po (Tidone, Trebbia, Nure, Arda, Taro, Parma, Enza, Secchia, Panaro), or directly into the Adriatic Sea (Reno, Lamone, Savio).

The climate of Emilia-Romagna has sub-continental characteristics, with cold winters and hot summers, moderated, however, by sea breezes along the Adriatic, while temperatures are closely affected by altitude in the Apennine region. Rainfall is about 800 mm/year and while evapotranspiration is around 500 mm.

With regards to population distribution, two zones are easily distinguished: the hills and mountains, thinly populated, and less suitable for economic development,

Table 3.8 Water withdrawals in Emilia-Romagna in 2001 (hm^3)

	Civil Mun.	Industries	Agriculture	Total
Ground water	279	169	212	660
Surface water	205	52	997	1254
Total	484	221	1209	1914

and the plains, characterized by an excellent communication network, possibilities for intensive farming and ideal conditions for industrial development.

Daily water consumption for domestic uses at a regional level is estimated in 158 l/cap with a value of 150 l/cap for the western provinces and a value over 160 l/cap for the eastern ones.

One of the problems in water supply concerns agricultural uses. Part of the surface water used for field irrigation, about 260 hm^3, is currently abstracted from the Apennine's rivers and torrents. This amount is expected to be dramatically reduced with the future application of the minimum vital discharge (DVM), which has to be assured in order to reach and keep a certain environmental quality level of the fluvial ecosystem in terms of river basin morphology, aquifer interaction, water quality, and hydrology. At present the Basin Authority of the Po River is studying and testing methods to compute the DVM, which has not been applied yet. Consequently, hypotheses in the usage of water at the regional level have been made, which reveal that lower water availability from hill and mountain basins (about 70 hm^3/year), will probably be covered by an additional 30 hm^3/yr abstracted from the aquifers, a water resource already overexploited. The greatest problems will appear in the Emilia region, and in particular in the provinces of Parma and Piacenza where the use of Apennine basin water is higher.

The exploitation of the groundwater resources in Emilia-Romagna is presently not sustainable in that the yield produces a water deficit. In general the aquifer is overexploited and this exacerbates the subsidence phenomenon and seawater intrusion along the coasts. The deficit in groundwater use represents the amount of water that exceeds the recharging capacity of the aquifer.

The shallow groundwater under Emilia-Romagna present extremely small deficits (fractions on hm^3/year) as far as the abstraction of Ferrara, Forlì-Cesena and Rimini provinces is concerned. Another contributing factor is that the main water

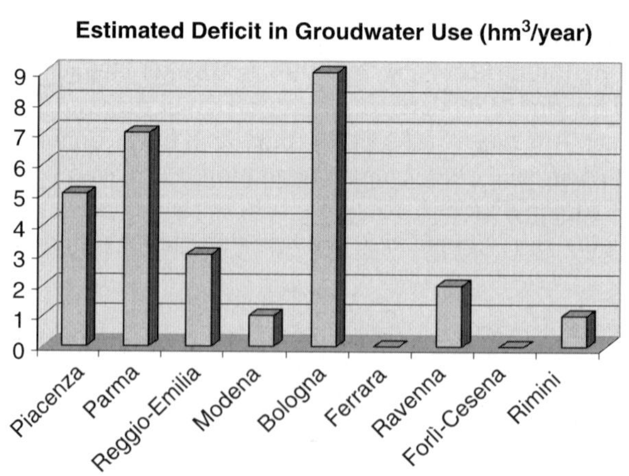

Fig. 3.8 Estimated deficit in groundwater use in Emilia - Romagna

source in Romagna is given as surface water by the Ridracoli Reservoir managed by Consortium Romagna Acque. The deficit is limited to 1–2 hm^3/year for the provinces of Ravenna and Modena and relevant for Bologna and Parma with 10–15 hm^3/year. Total regional deficit in underground water at annual level is about 30–35 hm^3/year.

River water quality can be considered adequate, at least for the stretches in the Apennine, where water can be directly supplied to domestic use without treatment. The use for irrigation is almost always permitted even though the stretches providing the provinces of Ferrara and Rimini have a high concentration of chlorides, requiring, thus, a good drainage of the cultivated soil. Generally, the qualitative characteristics of river stretches in the plains do not permit direct drinking use, neither favour the aquatic life of plants and animals.

The main problem in groundwater quality lies in the presence of nitrates, mostly in the areas under alluvial cones where their concentration is greater than 50 mg/l (limit for drinking water consumption). Therefore expensive treatment or mixing with water of better quality is required. However, the Groundwater Quality Status Index, used to study the distribution of antropogenic pollutants and natural chemical parameters, shows that Emilia-Romagna groundwater belong to the “class 0” of the Index Range of values, denoting null or insignificant impact of human activities and medium-good natural hydro-chemical parameters.

Coastal water quality can be classified as “medium”. The eutrophication level denotes presence of nutrients and algal biomass associated to low transparent waters and suffering benthic life ecosystem. On the other hand, a study conducted in 1999 pointed out that bathing and swimming are allowed in the 99.7% of the controlled coasts, thus denoting a low presence of urban pollution loads.

The problem of eutrophication is not limited to the Adriatic coast but affects the entire Po River Basin, as declared by the European Commission within the Urban Waste Water Treatment Directive 91/271/EEC. The role of agriculture in Emilia-Romagna will be determinant as the regional and local authorities are studying plans to re-use treated wastewater for irrigation: this would have the advantages of eliminating the discharge of treated wastewater in the Po River, and reducing groundwater withdrawals.

Prices are set by local authorities according to general guidelines. In Bologna District for example, services are provided by the public firm Seabo and a typical

Table 3.9 Qualitative chemical classes of groundwater

Class	Human Impact	Natural Hydro-chemical parameters
0	Null or insignificant	Medium–good
1	Null or insignificant	Optimum
2	Small and sustainable over the long period	Good
3	Significant	Medium
4	Remarkable	Bad

household with a consumption of 130 m³/year would pay approximately €132. Unit prices include wastewater treatment costs, about €0.276 /m³, water supply costs, €0.77 /m³ (for annual consumption between 80 and 150 m³), and sewerage services costs, €0.096 /m³.

Apart from Bologna and the other major urban centres, tourism in Emilia-Romagna is principally concentrated on the Adriatic coast. There are about thirty famous seaside resorts from the Comacchio Valleys to the Marches boundary. The beaches of Romagna, such as Milano Marittima, Cervia, Cesenatico, Bellaria, Rimini, Riccione, and Cattolica have, in fact, always attracted tourists from Italy and abroad.

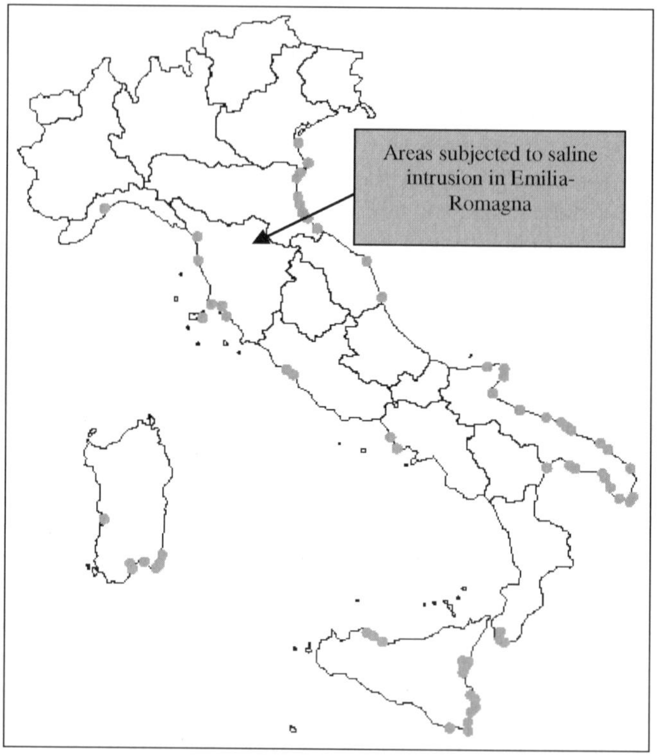

Fig. 3.9 Areas with saline intrusion phenomenon in Italy (source: Enea 1998)

Table 3.10 Regional tourist presence in Emilia-Romagna in 2001

Area	2000	2001	Trend% (2001–2000)	Diff.
Adriatic Coast	39,475,000	40,690,000	+3.1%	+1,125,000
Apennine	2,812,000	2,835,000	+0.8%	+23,000
Cities	3,403,000	3,480,000	+2.3%	+77,000
Watering places	1,994,000	2,025,000	+1.6%	+31,000
Total	47,684,000	49,030,000	+2.8%	+1,346,000

Table 3.11 Tourist presence in Adriatic coast in 2001

Nation	2000	2001	Trend% (2001–2000)	Diff.
Italy	31,642,000	32,480,000	+2.6%	+838,000
Germany	3,505,000	3,724,000	+6.2%	+219,000
Switzerland	662,000	687,000	+3.8%	+25,000
Austria	316,000	333,000	+5.4%	+17,000
Others	4,483,000	4,744,000	+5.8%	+261,000
Total	39,475,000	40,690,000	+3.1%	+1,215,000

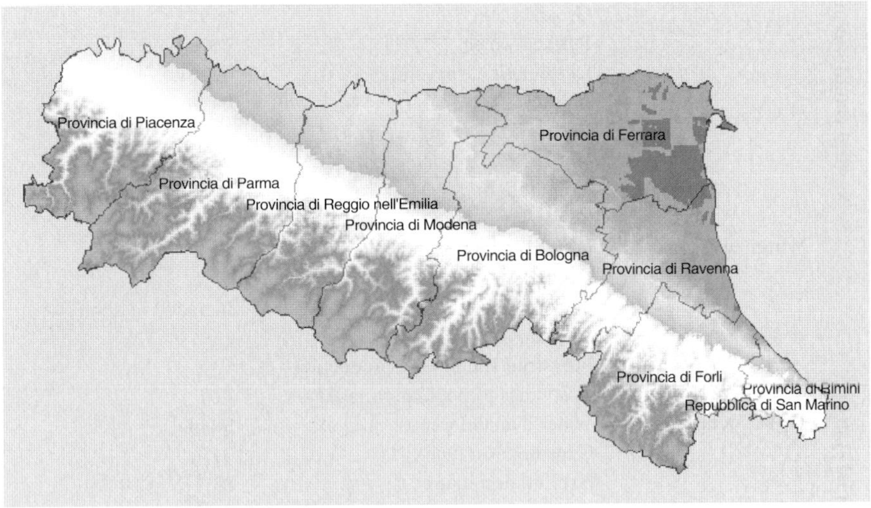

Fig. 3.10 Administrative provinces' division in Emilia-Romagna

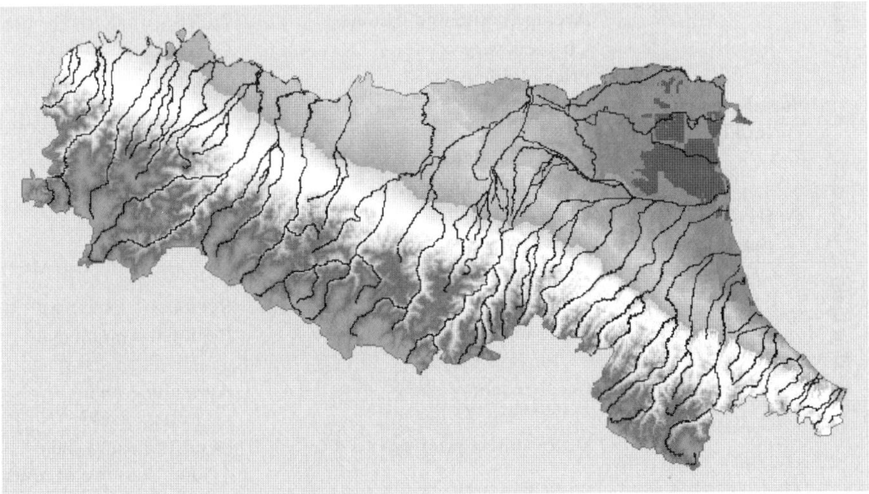

Fig. 3.11 Rivers of Emilia-Romagna

Table 3.12 Emilia-Romagna matrix of circumstances

Natural conditions and infrastructure	**Regional context**	Climate type	Subcontinental
		Aridity Index	1.6
		Permanent population	3,924,456
		Area	22,123
	Water availability	Total water resources/availability	1,925
		Trans-boundary water	0 (No)
	Water quality	Quality of surface water	Medium
		Quality of groundwater	Good
		Quality of coastal water	Medium
	Water supply*	Percentage of supply coming from:	
		Groundwater	24%
		Surface water	76%
		Desalination, recycling	0 (No)
		Importing	0 (No)
		Network coverage:	
		Domestic	94.8%
		Irrigation	100%
		Sewerage	71%
Economic and social system	**Water use***	Water consumption by category:	
		Domestic	15%
		Tourism	Included in domestic
		Irrigation	32%
		Industrial and energy production	53%
		Resources to population index	720
	Water demand	Water demand trends	Stable
		Consumption index	64%
		Exploitation index	83%
	Pricing system	Average household budget for domestic water (pa)	0.65%
		Average household budget for agricultural water	2.5%
		Average household income	20228
		Cost recovery	Good
		Price elasticity	Medium
Decision making process	**Social capacity building**	Public participation in decisions	Average
		Public education on water conservation issues	Average
	Water resource management	Water ownership	Public
		Decision making level regarding:	
		Water supply for each sector	Regional – municipal
		Water resources allocation for each sector	Regional – municipal
	Water policy	Local economy basis	Agriculture and tertiary sector
		Development priorities	Demand management and Wastewater re-use

* Water supply and water use computed from 1997–2000 data and power generation counts for 900 Hm^3

3.6 The Belice Basin, Italy

The Belice Basin is placed in the south-west of Sicily. It covers an area of about 967 km^2, in the administrative territory of Palermo, Trapani and Agrigento provinces. It borders with the Modione and Freddo River Basins at the west, with those of Jato and Oreto at the north and with those of Verdura and Carboj at the east. The Belice River is divided into three branches, the Right Branch, the Left Branch and the stretch after the confluence near the town of Poggioreale, each one defining sub-basins. The Right Branch has a length of 55 km and comes from the northern part of the basin, covering an area of 227 km^2. The Left Branch has a length of 57 km and comes from Mount Leardo and Mount Rocca Busambra and is supplied by the torrents Fosso and Bicchinello. Some of its tributaries are Corleone River and the torrents of Batticano and Realbate. The sub-basin has an area of 407 m^2. After the confluence, the river extends for 50 km up to the Sicily Canal.

In the Belice Basin there are many aquifers which are highly exploited for irrigation and municipal supplies.

The mean daily precipitation and evapotranspiration are respectively about 1.256 mm and 3.23 mm, yielding an Aridity Index of 0.39. Therefore, the region is characterized as semi-arid.

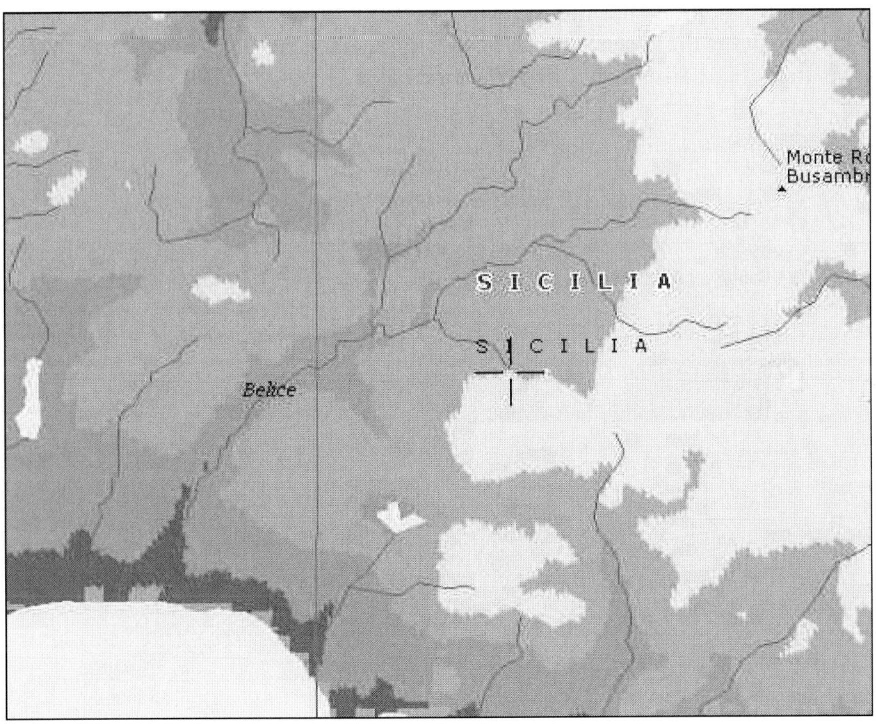

Fig. 3.12 The Belice Basin

The total water availability is 80 hm^3 of which 24.8 are groundwater. The water sources are groundwater, (19%), and surface water from Belice River and artificial lakes (81%). Corleone city imports water from Prizzi Lake at a rate of 18 l/sec, which corresponds to approximately 0.56 hm^3/year. About 21.9 hm^3/year are transferred to other neighbouring basins.

Groundwater is used mainly for drinking water provision for municipalities (45% of the total), while 30% is used for irrigation and 25% for industrial use.

Water consumed for irrigation is the 64% of the total, reflecting the fact that agriculture is at the basis of the local water economy, while domestic consumption, including tourism, is about 27%. Consumption for industrial and energy production is 9%.

The water quality of the Belice River, of groundwater and of coastal water is good.

The Belice Basin extends under the administration of Consortium N°3 Agrigento. Under this territory, the lack of water for irrigation in 2001, mainly due to the lower precipitation, caused an important delay of the irrigation season from the usual and official starting month of April to the month of June. Additionally, field requirements for the entire region were not completely covered due to uneven water availability conditions of the different water reservoirs, and consequently for some districts the irrigation season was terminated one month earlier.

Natural conditions are not the only problem affecting irrigation water needs: in some districts the absence of the necessary maintenance of pipelines transferring water from the reservoirs has been the first cause of pipe bursts.

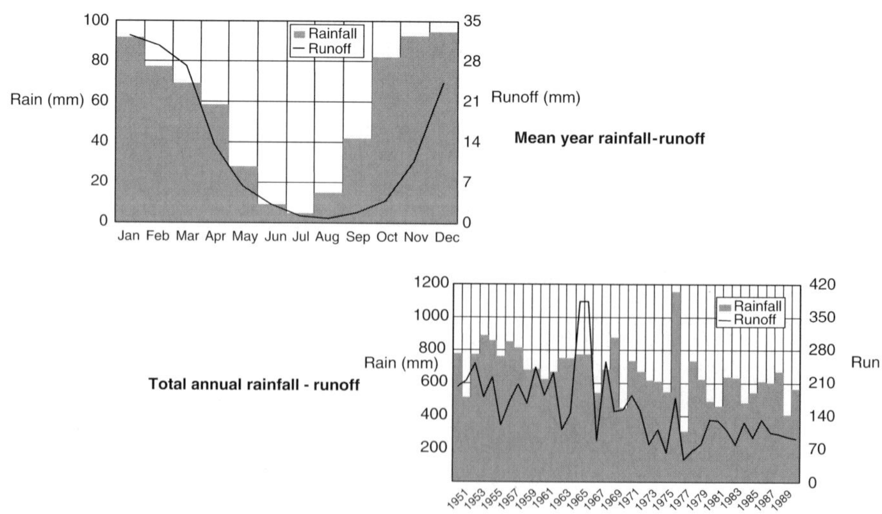

Fig. 3.13 Rainfall and runoff in the Belice Basin

Table 3.13 Belice basin matrix of circumstances

Natural conditions and infrastructure	**Regional context**	Climate type	Warm-temperate
		Aridity Index	0.38
		Permanent population	55,329
		Area	967
	Water availability	Total water resources/availability	80
		Trans-boundary water	0.56
	Water quality	Quality of surface water	Good
		Quality of groundwater	Good
		Quality of coastal water	Good
	Water supply	Percentage of supply coming from:	
		Groundwater	19%
		Surface water	81%
		Desalination, recycling	0 (No)
		Importing	0 (No)
		Network coverage:	
		Domestic	100%
		Irrigation	100%
		Sewerage	100%
Economic and social system	**Water use**	Water consumption by category:	
		Domestic	27%
		Tourism	(inc. in Domestic)
		Irrigation	64%
		Industrial and energy production	9%
		Resources to Population index	390
	Water demand	Water demand trends	Increasing
		Consumption index	100%
		Exploitation index	100%
	Pricing system	Average household budget for domestic water (pa)	0.47%
		Average household budget for agricultural water	5%
		Average household income	16740
		Cost recovery	Good
		Price elasticity	Medium
	Social capacity building	Public participation in decisions	Bad
		Public education on water conservation issues	Bad
Decision making process	**Water resource management**	Water ownership	National (public)
		Decision making level (municipal regional national) regarding:	
		Water supply for each sector	Regional
		Water resources allocation for each sector	Regional
	Water policy	Local economy basis	Agriculture, tourism
		Development priorities	Agriculture, tourism

3.7 Tel Aviv Region, Israel

The region is located in the coastal plain on the eastern shore of the Mediterranean Sea and it lies above the coastal aquifer. In terms of population, the Tel Aviv region is the largest in Israel with two million people, 30% of the total population (Table 3.15).[3] The region has 160,000 dunam of cultivated agricultural land, 5% of the total cultivated land in the country. The region's water economy is therefore characterized by relatively high domestic and industrial consumption, and relatively low agricultural consumption.

Natural water sources in the area are:

- Supply from the national water system (via the national water network of the Mekorot company)
- Production from the coastal aquifer, above which the region lies
- Water supply from the Sea of Galilee via the National Water Carrier (NWC)

In addition, part of the fresh water is provided by private producers from the coastal aquifer (some 35% of fresh water).

In the future, this region is slated to receive a significant amount of the desalinated seawater.

Aggregate supply is summarized in Table 3.20. Domestic consumption is similar to the national average (100 m^3/capital/year), and expected to increase by 20% with the development of metropolitan parks and the improvement in quality of life.

The quality of the freshwater is good, with a salinity level of 150–250 mg chloride per litre. In the future, the use of desalinated water will lead to an improvement in the water quality. The region's large population creates the potential for a large supply of recycled water for agriculture. In addition, high quality treated wastewater can be used for irrigation of metropolitan parks and for rehabilitation of streams like the Yarkon River.

The region is characterized by a Mediterranean, semi-arid climate. The annual precipitation is 450 mm and the aridity index varies from 0.05 to 0.2.

Water prices are determined within the national framework. Private producers are subject to a production levy. Land prices in this area are among the highest in the country, and therefore the region is subject to further urbanization and a reduction in agricultural area. Agriculture in this region has value as a public good in conserving open areas and "green lungs".

3.8 Arava Region, Israel

The region is located at the south-eastern tip of Israel, between the Dead Sea and the Red Sea.

[3] Data in all tables refers to the year 2000.

Table 3.14 Tel-aviv matrix of circumstances

Natural conditions and infrastructure	**Regional context**	Climate type	Semi-arid
		Aridity Index	0.05-0.2
		Permanent population	1,883,700
	Water availability	Total water resources / Availability	338
		Trans-boundary water	
	Water quality	Quality of surface water	–
		Quality of groundwater	Good
		Quality of coastal water	–
	Water Supply	Percentage of supply coming from:	
		Groundwater	97%
		Surface water	
		Desalination	0%
		Recycling	3%
		Importing	0%
		Network coverage:	
		Domestic	100%
		Irrigation	100%
		Sewerage	100%
	Water use	Water consumption by category:	
		Domestic	56%
		Irrigation	17%
		Industrial and energy production	27%
		Population to resources index	
	Water demand	Water demand trends	
		Domestic	Steadily increasing
		Industrial	Steadily increasing
		Agriculture	Transfer to recycled water
		Rivers	Use of recycled water
		Consumption index	Stable per-capita urban cons.
		Exploitation index	100%
Economic and Social System	**Pricing system**	Average household budget for domestic water (pa)	$100
		Average household budget for agricultural water	
		Average household income	
		Cost recovery	
		Price elasticity	Agriculture – somewhat elastic. Urban and industrial - small.
	Social capacity building	Public participation in decisions	Very high
		Public education	Fair
Decision making -process	**Water resources management**	Water ownership	State
		Decision-making level (municipal, regional, national) regarding:	
		Water supply for each sector	National
		Water resources allocation for each sector	National
	Water policy	Local economy basis	National
		Development priorities	Recycling / desalination

Table 3.15 Distribution of population by type of settlement

Type of settlement	Tel - Aviv		Arava	
	thousands	%	thousands	%
Metropolitan areas (Pop. exceeding 200,000)	349	19	–	0
Big cities (Pop. 100,000–200,000)	887	47	–	0
Mid-sized cities (Pop. 20,000–100,000)	469	25	40	88
Small towns and cities (Pop. 2,000–20,000)	154	8	–	0
Villages and communities	25	1	5	12
Total	1,884	100%	45	100%

Table 3.16 Domestic consumption (MCM/year)

Year	m^3/capita	Consumption from national system	Consumption from local system	Total demand
Tel - Aviv	100	75	113	188
Arava	200	–	9	9

Table 3.17 Industrial consumption (MCM/year)

	Freshwater					
	Consumption from national system	Consumption from local system	Total demand	Saline water	Recycled water	Total
Tel - Aviv	23	35	58	0	0	58
Arava	0	1	1	0	0	1

Table 3.18 Agricultural consumption (MCM/year) sources

		Tel - Aviv	Arava
National System	Fresh	34	–
	Recycled	5	–
	Saline	–	–
	Total	39	–
Local System	Fresh	51	12
	Recycled	–	5
	Saline	–	14
	Total	51	31
TOTAL	Fresh	85	12
	Recycled	5	5
	Saline	–	14
	Total	90	31

Table 3.19 Environmental consumption (MCM/year) sources

		Tel - Aviv	Arava
Local System	Fresh		
	Recycled	2	
	Saline		
	Total	2	

Table 3.20 Summary of water consumption by water type (MCM/year)

		Tel – Aviv	Arava
National system	Fresh	133	–
	Recycled	5	–
	Saline	–	–
	Total	138	–
Local System	Fresh	198	22
	Recycled	–	5
	Saline	–	14
	Total	198	41
Total	Fresh	331	22
	Recycled	5	5
	Saline	–	14
	Total	336	41

Table 3.21 Salinity levels and long-term average recharge by water resource

Basin	Salination level (mgchlorine/liter)	Average annual-recharge (MCM)
Coastal Aquifer – national system and local producers	–	250
Sea of Galilee Basin – national system	–	180
Arava – local sources	400	–
Total	400	222

Table 3.22 General water balance (MCM/year) + estimates

	Arava	Tel – Aviv
Demand by sector:		
Domestic	9	188
Industrial	1	58
Agricultural	31	90
Jordan & PA	–	–
Environment	–	2
Total	41	338
Demand by water type:		
Freshwater	22	331
Reclaimed	5	7
Saline	14	–
Total	41	338
Supply:		
Aquifers (including saline)	27	333
Desalination	9	–
Recycled	5	5
Total	41	338

The region is sparsely populated, based mainly on the tourist city of Eilat, at the southern tip. The remaining population is scattered in rural villages.

Domestic consumption per capita in this region is particularly high, because of dry climatic conditions, that lead to heavy evaporation and a higher demand for garden irrigation and drinking water. In addition to this, a large part of the population lives in rural settlements, where large amounts of water are needed for private and public gardens.

The sources of fresh water in the region are the following:

- The Arava is not part of the national water system, but receives water from local sources only, via the national water company Mekorot.
- Drillings in the centre of the region (Faran drillings) yield water of reasonable quality: up to 350 mg chlorine per litre.
- Drillings in the southern Arava yield low-quality water: 600–1,100 mg chloride per litre.
- The desalination plant of Red Sea water provides water for the local population in Eilat.

In addition, waste water for agriculture is obtained from Eilat and the agricultural settlements. It is important to note that the Red Sea is a unique coral reserve of great ecological value, and it is therefore essential that wastewater is recycled for agriculture and not discharged in the sea.

The prices for all water supplied by Mekorot, fresh and saline, are determined within the national framework. Saline water is cheaper than fresh water, in accordance with the salinity level. The price for recycled water for agriculture covers the operational and the capital costs, after discounting state grants.

Water development plans for the region focus on pooling and transferring wastewater. In the more distant future there is a possibility that the desalination plant in Eilat will be enlarged.

The region is characterized by an arid climate with a very low precipitation (rainfall up to 10 mm/year) and the aridity index is 0.65.

The climatic conditions favour intensive cultivation of vegetables, flowers and date palms. Some 40% of the greenhouses in Israel are located in this region.

In addition to the above, this region borders with Jordan. The water production balance – drillings and water production from the local aquifer – is affected by the peace treaty with Jordan. Land prices are low and there is no demand for additional urbanization.

3.9 Akrotiri Region, Cyprus

The Akrotiri area covers the Akrotiri peninsula and it is the southernmost part of the island, covering an area of 142 km^2. Its eastern part is taken over by the urban area of Limassol with some 125,000 inhabitants. There are 10 other village communities with a total population of 16,000 within the same area that basically may be

Table 3.23 Arava matrix of circumstances

Natural conditions and infrastructure	**Regional context**	Climate type	Hyper-arid
		Aridity Index	0.5–0.65
		Permanent population	45,200
	Water availability	Total water Resources/ Availability (MCM)	41
		Trans-boundary water	
	Water quality	Quality of surface water	–
		Quality of groundwater	Poor
		Quality of coastal water	–
	Water supply	Percentage of supply coming from:	
		Groundwater	84%
		Surface water	0%
		Desalination	16%
		Recycling	0%
		Importing	
		Network coverage:	
		Domestic	100%
		Irrigation	100%
		Sewerage	100%
	Water use	Water consumption by category:	
		Domestic	22%
		Irrigation	76%
		Industrial and energy production	2%
		Population to resources index	
Economic and social system	**Water demand**	Water demand trends	
		Domestic	Stable
		Industrial	Stable
		Agriculture	Transfer to recycled water
		Rivers	–
		Consumption index	Stable per-capita urban consumption
		Exploitation index	100%
	Pricing system	Average household budget for domestic water (pa)	$130
		Average household budget for agricultural water	
		Average household income	
		Cost recovery	
		Price elasticity	Agricultural - somewhat elastic. Urban and industrial - small.
	Social capacity building	Public participation in decisions	Very high
		Public education	Fair

(continued)

Table 3.23 (continued)

Decision-making process	**Water resources management**	Water ownership	State
		Decision making level (municipal, regional, national) regarding:	
		Water supply for each sector	Local
		Water resources allocation for each sector	Local
	Water policy	Local economy basis	National
		Development priorities	Recycling and desalination

considered as suburbs to Limassol with their inhabitants commuting to the city and also working as farmers within the overall area.

At the southern tip of the peninsula there is a major British military base with an airfield and an estimated force of the order of 15,000 soldiers with their families. This is separated from the agricultural land and aquifer further to the north, by a Salt Lake and marshland that is of unique environmental importance.

The aquifer, the third largest in the island, is essentially a gently dipping coastal deltaic alluvial aquifer of a 40 km^2 extent. Its western half coincides with the alluvial fan deposits of the Kouris River that drains a catchment of 338 km^2, while the Garyllis River, draining a watershed of 100 km^2, takes up the eastern half. Groundwater is pumped through some 500 wells and boreholes mainly for irrigation (9 to 12 hm^3/year) and for domestic purposes (1.5–3 hm^3/year). About 90% of the annual extraction is metered and recorded at monthly intervals. Pumping permits are issued annually on the basis of the current groundwater conditions and the water stored in the surface reservoirs.

The main source of the natural recharge of the aquifer, after the construction of Polemidhia dam in 1965 on Garyllis River and the construction of the Kouris dam in 1987 on Kouris River, changed dramatically. It now depends entirely on local rainfall (about 380 to 430 mm), on return flow from water imported for irrigation and on artificial groundwater recharge.

Sea intrusion was originally confined at the eastern part. More recently, and after the construction of the Kouris dam of 115 hm^3, a large part of the Kouris delta area has also been sea-intruded. Ongoing artificial groundwater recharge with water from surface reservoirs, and planned with treated effluent, together with further control of pumping is expected to reverse the situation. Furthermore, with the reduction of the flashing effect of the annual recharge together with the increased agricultural activity, a gradual build-up of nitrate and other elements has been noted in the groundwater of the area.

Presently the irrigation requirements of the area are met by local groundwater, tertiary treated effluent, surface water from the Kouris, Germasogeia and Polemidhia dams, and by reclaimed groundwater pumped from the Limassol urban area. The area irrigated at present is 2,200 ha with a demand of 15 hm^3, out of a planned area of 3,775 ha. The Limassol domestic supply is mainly provided from the Kouris dam and partly from groundwater from the Germasogeia aquifer.

The aquifer is very well controlled. The groundwater levels are observed monthly from a network of 150 since 1960, 85 to 100 of which are regularly sampled. The groundwater pumping is quite well monitored through water meters that are observed every month. A good database exists and numerous studies have been performed including groundwater modelling.

Low rainfall and reduction of the surface reservoir water content resulted to diminished recharge, both natural and artificial, of the aquifer. This together with the continued extraction pattern of pre-dam construction has caused a serious drop of the groundwater levels (Fig. 3.14 to Fig. 3.16) and sea intrusion (Fig. 3.17).

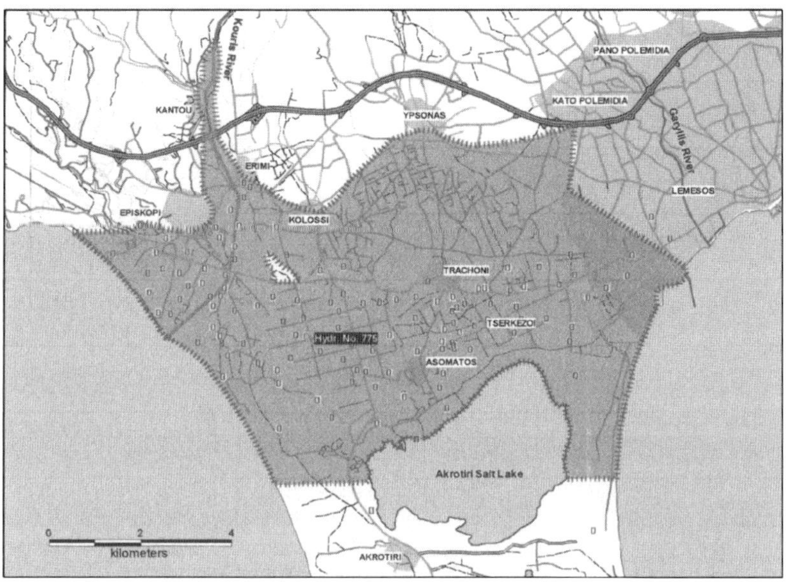

Fig. 3.14 Akrotiri aquifer showing inhabited areas and well observation network

The intensive use of fertilizers in agriculture together with the reduction of the flashing effect by natural recharge resulted to a nitrate built-up in the aquifer. The concentration of nitrate ion in the eastern part of the aquifer is in excess of 200 mg/l.

At the same time the diminished flows of Kouris River and the drop of groundwater levels are threatening the ecosystem of the marshlands and that of the Salt Lake. Urbanization at the eastern and northern parts of the aquifer and increased storm runoff from these areas create a problem to the Salt Lake since it is on the lowland, being, thus, the natural receiving area.

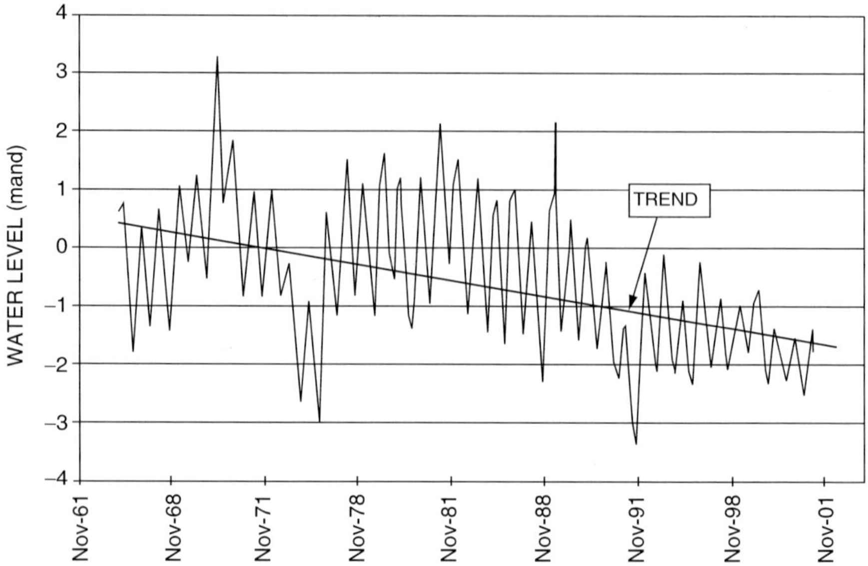

Fig. 3.15 Hydrograph of borehole Akrotiri 775 (Elev. 15.63 m amsl)

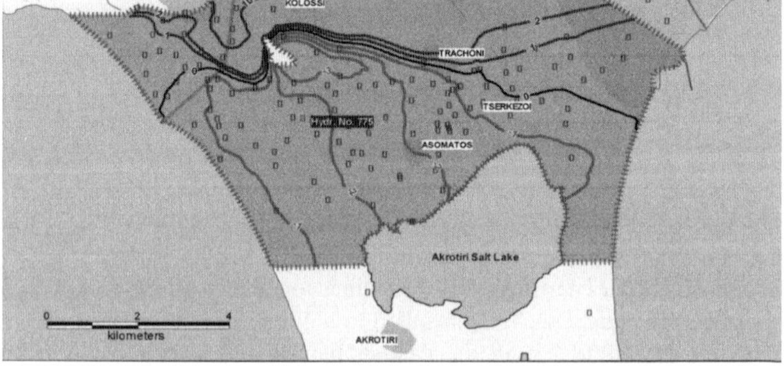

Fig. 3.16 Akrotiri aquifer - Water level (m amsl) contour map March 2001

Human interventions have changed dramatically the hydrologic regime in the area, especially after the construction of the Kouris dam. The average water balance over the period of 1967/68 to 1976/77 compared to present conditions is shown on the Table below (All in hm^3/year).

Table 3.24 Recharge

Kouris dam was constructed in 1987	Rainfall	Riverbed Recharge	Subsurface Inflow	Sea intrusion	Return from irrigation	Return imported/ Diversions	Artificial Recharge	Total	Remarks
Before (1968–1978)	5.9	15.4	4.2	0.7	4.5	3.5		34.2	Average rainfall 395 mm
At present	4.2	0.5	0.2	3.0	1.1	0.7	3.3	13.0	Average rainfall 380 mm

Table 3.25 Outflow

	Abstraction for irrigation and domestic	Evapotran-spiration	Rising water	Sea/ lake outflow	Total	Remarks
Before (1968–1978)	14.5	2.5	2.2	16.0	35.2	
At present	10.8	2.4	0.3	0.5	14.0	

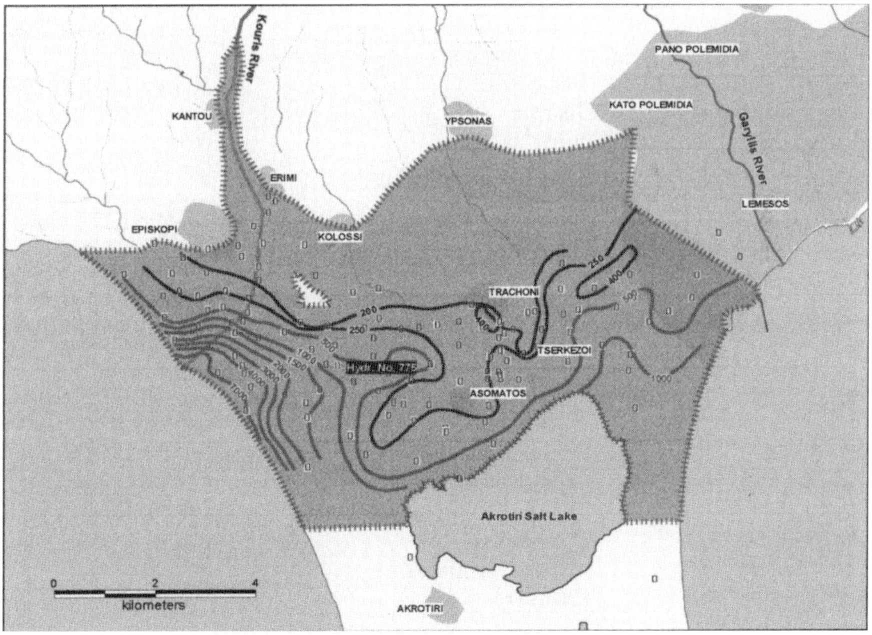

Fig. 3.17 Akrotiri aquifer - Isochloride (ppm) contour map April 2001

From the balance above, one should note the very small quantity of groundwater that outflows from the system at present and which does not provide the required leaching effect, the reduction of rising water that affects the marshland, and the increase in sea intrusion quantities which although they are considered as part of the "recharge", the resulting true balance is in effect negative by an order of $4\,hm^3$ per year.

Table 3.26 Akrotiri area matrix of circumstances

Natural conditions and infrastructure	**Regional context**	Climate type	Csa-Mediterranean
		Aridity* Index	Semi-arid (0.330)
		Permanent population	156000
	Water availability	Total water resources / Availability	$30\,hm^3$
		Trans-boundary water	–
	Water quality	Quality of surface water	Very Good
		Quality of groundwater	Fair - Poor
		Quality of coastal water	Good
	Water supply	Percentage of supply coming from:	
		Groundwater	33%
		Surface water	62%
		Desalination/Recycling	5%
		Importing	
		Network coverage:	
		Domestic	100%
		Irrigation	>85%
		Sewerage	Apx. 75%
Economic and Social issues	**Water use**	Water consumption by category:	
		Domestic	30%
		Tourism	10%
		Irrigation	60%
		Industrial and energy production	
		Resources to population index	$192\,m^3/c$
	Water demand	Water demand trends	Increasing
		Consumption index	100% Increasing
		Exploitation index	100% Increasing
	Pricing system	Average household budget for domestic water**	€99.2/yr
		Average household budget for agricultural water	€0.11/m^3
		Average household income**	€24,207urban €18,488 rural
		Cost recovery	Dom €0.58/m^3, Irr €0.11
		Price elasticity	Very small

(continued)

Table 3.26 (continued)

Decision-making process	**Social capacity building**	Public participation in decisions	Fair
		Public education on water conservation issues	Fair
	Water resources management	Water ownership	State – (partly private)
		Decision-making level regarding:	
		Water supply for each sector	National
		Water resources allocation for each Sector	National
	Water policy	Local economy basis	Agri/tertiary
		Development priorities	Agri/tourism

*Aridity = 407/ (4.5×365×.0.75) 1961–1990

**Family Budget Survey 1996/97 Statistical Service of Republic of Cyprus

3.10 Germasogeia Region, Cyprus

The Germasogeia catchment is in the southern coast of Cyprus. It is about 141 km^2 up to the Germasogeia dam of 13.1 hm^3 capacity. Its average annual flow is about 20 hm^3. A major part of the catchment is covered by natural forest but considerable agricultural activity is present in riparian land. The annual and seasonal crops irrigated from the various sources of water are shown on Table 3.27.

There are 14 village communities within the watershed with a total permanent population of just over 10,000 and water demand of 0.5–0.7 hm^3 per year, of which 12 villages are upstream the dam with a population of about 4,000. There is considerable tourist development at the coastal area with an estimated 0.5 million-guest nights and a water demand of 0.9 hm^3 during the tourist season.

Downstream the dam a typical river alluvial aquifer develops. This aquifer, which is 5 km east of Limassol town (Fig. 3.18), has a length of 5.5 km and an average width of about 350 m. This phreatic aquifer consists of sandy gravels with low silt content except towards the coast where an increase of finer material is noted. The thickness in the deepest part varies from 35 m near the dam to 50 m near the coast. The permeability in the upstream part of the aquifer is as high as 300 m/d reducing to 100 near the Delta. The specific yield varies from 13 to 22%. The active storage of fresh water is of an order of 3.5 hm^3 increasing to 5.0 hm^3 at high water table.

The small aquifer between the surface reservoir and up to 4 km downstream, before the development of the Delta area, has been relied upon to meet the major portion of the increasing demand for the water supply of the town of Limassol and neighbouring villages with high seasonal demand due to tourism.

Since the construction of the dam in 1968, the recharge of the aquifer depends on controlled releases from the dam and its spills. During the last ten years the dam spilled only twice, in 1993 and 1995. The complete cut-off of natural replenishment by the construction of the dam and the proximity to the sea, coupled with the

Table 3.27 Annually and seasonally irrigated crops (in hectares and hm^3)

Germasogeia Watershed (in relation to dam)	River		Springs		Wells/bh		Total	
	Area	Water use	Area	Water use	Area	Water use	Area	Water use
Upstream	295	2.5	78	0.8	16	0.2	386	3.5
Downstream	304	3.0					304	3.0

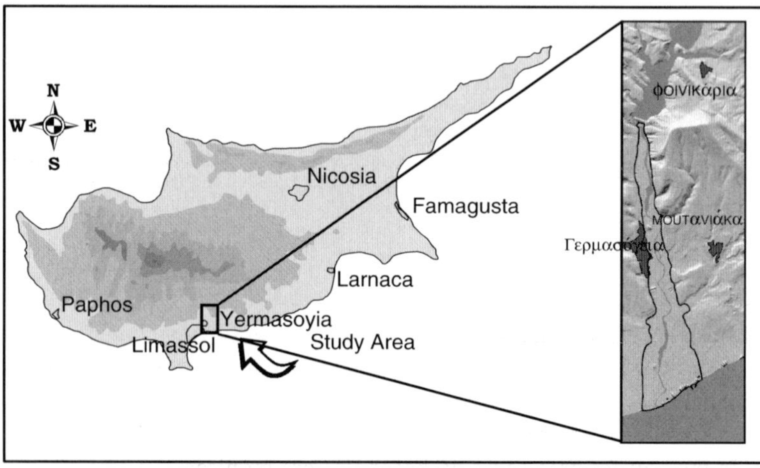

Fig. 3.18 Location of the Germasogeia watershed (the part downstream the dam)

increasing extraction from the aquifer, requires a coordinated programme of releases from the dam for artificial recharge to cope with the extraction. With such action the sea intrusion is controlled and at the same time an efficient use of the scarce water resources is made.

The need for controlled releases from the dam to artificially recharge this aquifer through flooding in the active channel became a necessity by 1982 due to the increasing demand for domestic supply and the rather dry conditions experienced at the time. This conjunctive use of the surface and groundwater reservoirs enabled a dramatic increase in the extraction from this aquifer deferring the need for an expensive treatment plant for many years.

The extraction was doubled due to an equivalent increase of recharge. It is important to note that with the regulated releases of water and the resulting recharge, the annual extraction in many years was about three times the active storage of the riverbed aquifer. In the beginning of the recharge, large quantities were released at irregular time intervals. Gradually, the daily release quantities were being reduced and the length of the period of release was increased.

Since 1986, the release is practically continuous and at such rates that the losses to the sea through the subsurface are minimal.

The rates of release are of the order of 15,000–25,000 m^3/d whilst the total groundwater inflow to the main well-field serving Limassol is in the range of 18,000 to 30,000 m^3/h.

Chemically, the groundwater is similar to that of the water in the surface reservoir. Bacteriological analyses from all the boreholes show that the 10–20 m of unsaturated thickness of alluvial sediments provide an efficient protection to bacteriological pollution.

In the Delta area and near the coast the interface has remained practically stable showing that the recharge-extraction regulation has been of the correct order without excessive pumping or serious subsurface loss of fresh water to the sea.

In effect the small Germasogeia riverbed aquifer has been turned into a natural treatment plant for domestic water supply without the need of complicated and expensive surface water treatment requiring chemicals, qualified technical and managerial personnel and the necessary civil engineering structures. Surface water from the Germasogeia and Kouris dams is being released in the riverbed since 1982 for recharge of the aquifer. Groundwater is pumped for the domestic water supply of the Limassol town, for the surrounding villages, and the tourist zone. This aquifer is the only source of domestic water supply of the local village communities and the tourist zone.

The catchment area has extensive hydrometeorological, geological and hydrogeological data as well sufficient surface and groundwater quality data. It constitutes an excellent case study for evaluating drought conditions and their repercussions on the hydrologic regime and the socio-economic environment of the area. In the aquifer area some 46 boreholes are monitored every 15 days and conductivity logs are kept for 10 boreholes for monitoring the sea/fresh water interface. The extraction from all wells and boreholes is monitored monthly by water-meters. The releases for recharge are monitored on a daily basis.

The Germasogeia water resources system (surface reservoir and aquifer) is the most intensively exploited one in the island. In 1996 up to 9 hm^3 of groundwater were extracted from this small aquifer, whose area is only 3 km^2 and its total fresh water capacity at average groundwater level conditions, is in the order of only 3.5 hm^3.

Table 3.28 Recharge

Germasogeia dam (13 hm^3) was constructed in 1968	Rainfall and return from irrigation /domestic	Riverbed recharge	Leakage from dam	Sea intrusion	Artificial recharge	Total	Remarks
(1982– 1987) average rainfall 430 mm	0.4	0.5	1.8	0.0	3.6	6.3	By A. Christodoulides
(1991–2000) average rainfall 400 mm	0.5		1.0	0.1	5.1	6.2	By A. Georgiou

Table 3.29 Outflow

	Abstraction for domestic	Sea outflow	Total	Remarks
(1982–1987) average rainfall 430 mm	5.6	0.7	6.3	By A. Christodoulides
(1991–2000) average rainfall 400 mm	6.4	0.3	6.7	By A. Georgiou

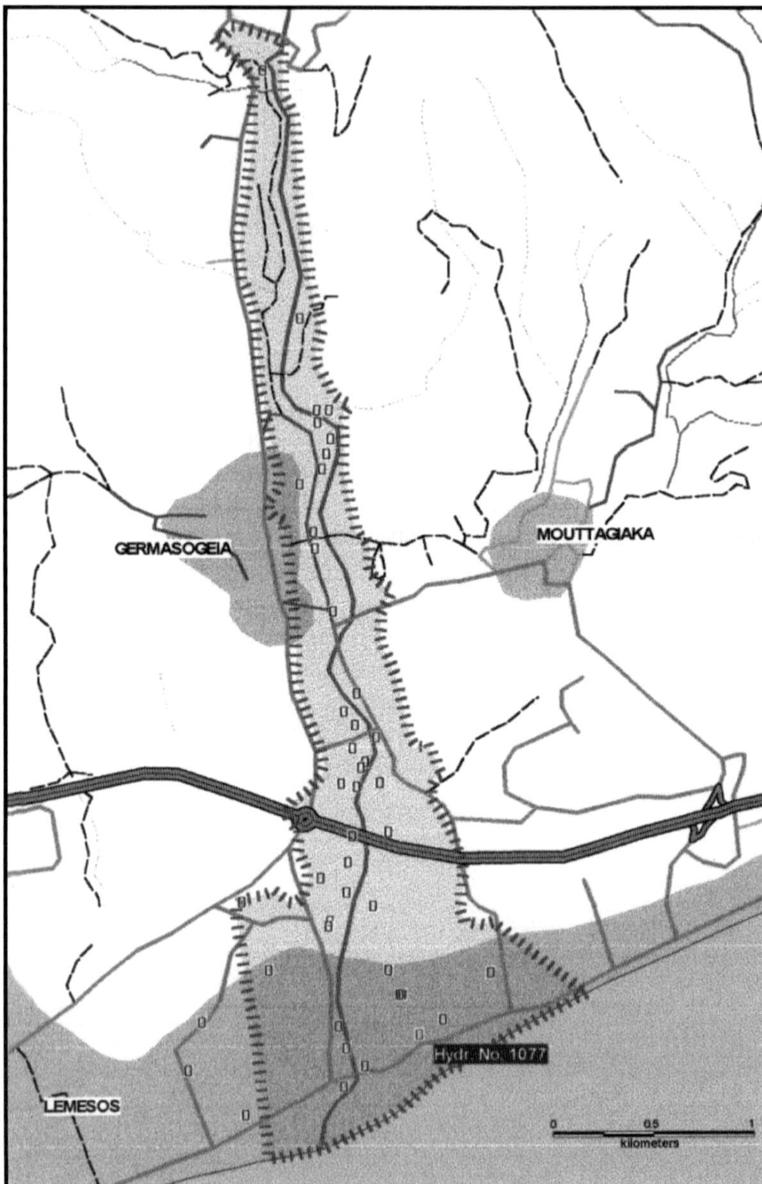

Fig. 3.19 Germasogeia riverbed aquifer - Location map

A fast-growing urbanization within the aquifer area and tourist development is causing concern about the environmental impact and possible deterioration of the quality of groundwater in this highly susceptible aquifer.

The hydrogeological regime and the water balance of the aquifer are "regulated" by controlled releases from the dam into the river valley. The main targets are:

- To cover water demand with groundwater of acceptable quality
- To protect the aquifer from sea intrusion
- To minimize groundwater losses to the sea
- To maximize the water availability through conjunctive use of surface and groundwater

Some 23 boreholes operate in the aquifer today for domestic water supply. The yields of these boreholes vary from 50 to 200 m3/h. Annually, the average extraction is about 6 hm^3, whilst the average artificial recharge is about 5 hm^3.

The water balance of the aquifer is quite good, and provided there is ample water in the surface reservoir for recharge and groundwater extraction does not exceed the capabilities of the system, there would be no problems of sea intrusion.

The sustainable extraction under natural conditions, i.e., with no artificial recharge of the aquifer is estimated to be of an order of 1.4 hm^3/year based mainly on the leakages from the dam.

Figure 3.19 shows the location of the surface reservoir and the aquifer whilst Fig. 3.20 shows a typical groundwater level fluctuation. The cycles of increased recharge or extraction are quite obvious on this hydrograph. Fig. 3.21 shows the groundwater contours in the area, especially as these develop around the main well-fields and near the coast.

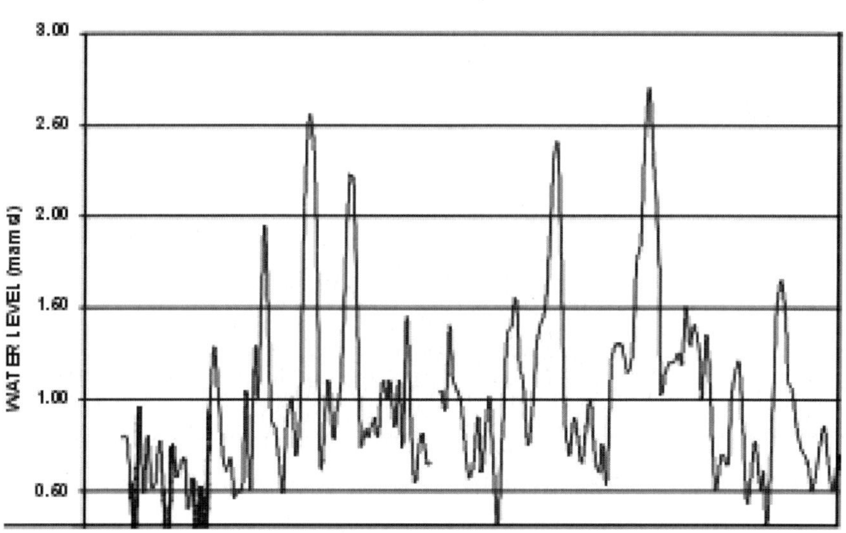

Fig. 3.20 Hydrograph of borehole Germasogeia 1077 (Elev. 4.98 m amsl)

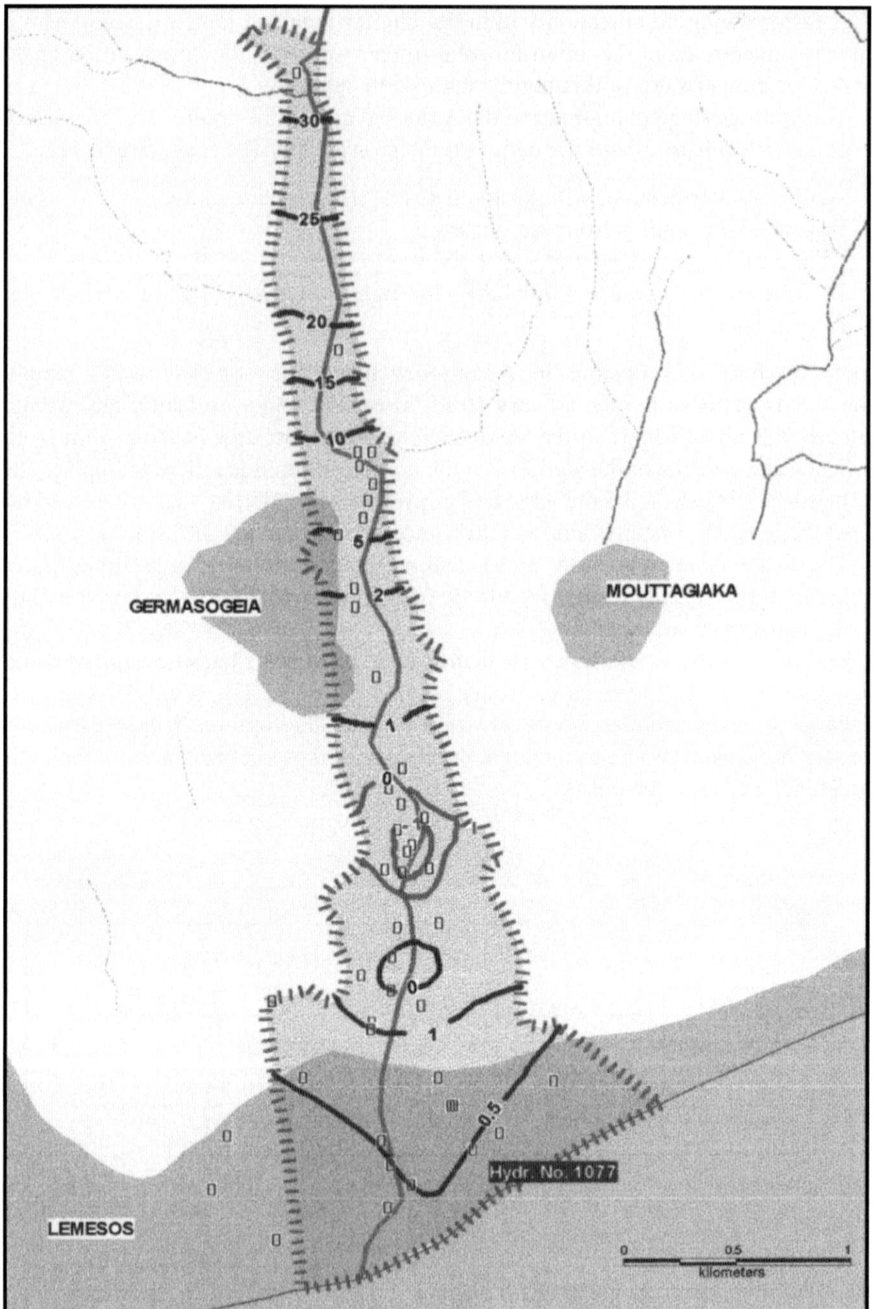

Fig. 3.21 Germasogeia riverbed aquifer water level (m amsl). Contour map November 2001

Table 3.30 Germasogeia area matrix of circumstances

Natural conditions and infrastructure	**Regional context**	Climate type	*Csa- Med/ean*
		Aridity* Index	*Semi-arid (0.356)*
		Permanent population	*10000*
	Water availability	Total water resources /availability	*20/12 hm^3*
		Trans-boundary water	–
	Water quality	Quality of surface water	*Very Good*
		Quality of groundwater	*Very Good*
		Quality of coastal water	Good
	Water supply	Percentage of supply coming from:	
		Groundwater	*15%*
		Surface water	*50%*
		Desalination, recycling	*35%*
		Importing/ Exporting	
		Network coverage:	
		Domestic	*100%*
		Irrigation	*>70%*
		Sewerage	*Apx. 80%*
	Water use	Water consumption by category:	
		Domestic	*0.6 hm^3*
		Tourism	*0.9 hm^3*
		Irrigation	*6.5 hm^3*
		Industrial and energy production	
		Resources to population index	*1,200 m^3/c*
Economic and social issues	**Water demand**	Water demand trends	*Increasing*
		Consumption index	*67% increasing*
		Exploitation index	*67% Increasing*
	Pricing system	Average household budget for domestic water**	*€99.2/yr*
		Average household budget for agricultural	*€ 0.11/m^3 depends on land*
		Water	
		Average household income**	*€ 24,207urban*
			€ 18,488 rural
		Cost recovery	*Dom € 0.58/m^3*
			Irr € 0.11
		Price elasticity	*Very small*
	Social capacity building	Public participation in decisions	*Fair*
		Public education on water conservation issues	*Fair*
Decision-making process	**Water resources management**	Water ownership	*State – (partly private)*
		Decision-making level (municipal, regional, national) regarding:	
		Water supply for each sector	*National*
		Water resources allocation for each Sector	*National*
	Water policy	Local economy basis	*Agri/tertiary*
		Development priorities	*Agri/tourism*

*Aridity = 478/ (4.9 × 365 × .0.75) 1961–1990

**Family Budget Survey 1996/97 Statistical Service of Republic of Cyprus

3.11 Kokkinochoria Region, Cyprus

The area includes five village communities and three municipalities with a total permanent population of 30,000 and an annual water demand in excess of 1.7 hm^3. Two areas near the coast (Paralimni and Agia Napa) have been developed into very attractive tourist resorts with tourists exceeding 6 million overnights and a water demand of about 3 hm^3.

The Kokkinochoria area, being at the lee-side and far from the Troodos Mountains, receives the lowest rainfall in the island, the long-term average being 330 mm/year. There is no stream crossing the area except during storm events in winter time due to local storms.

The local aquifer has been overexploited since the early 1960s and groundwater mined is in excess of 350 hm^3. At present, groundwater reserves are only 15% of the original. Water levels in the aquifer within 2 km from the coast have dropped to 50 m below mean sea level.

The region is an early-potato producing area with most of the production being exported to the UK and elsewhere. The past agricultural activity in the area has been maintained by importing water through the Southern Conveyor Project from the Kouris Dam which is situated about 70 km to the west. A total of an annual supply of 17 hm^3 has been envisaged which would allow the continuation of the agricultural activity in the area together with the local safe yield of 8 hm^3. This has been accomplished, although the extended drought of the last decade did not allow the transfer of the quantities envisaged. This did not have devastating repercussions since a lot of the workforce shifted in the meantime from agriculture to other employment associated with the locally thriving tourist industry. Nonetheless, both the soils and farming experience in the area is a resource that should be exploited to its maximum and the conditions need to be established in the area to allow the continuation of potato production for the benefit of the economy of the island.

There is satisfactory hydrogeological information on the area with some 164 wells being monitored every three months since 1964 as far as water levels are concerned. Water quality surveys are carried out seasonally to check the propagation of the sea-intrusion.

Table 3.31 shows an estimated water balance of the Kokkinochoria aquifer for two periods: 1963–78 and from 1990 to present.

The most productive parts of the aquifer (Ormidhia, Xylophagou, Liopetri, Phrenaros), have been sea intruded and abandoned since the early 1980s. The less productive parts have already been depleted with dramatically reduced borehole yield.

It is estimated that over 5,000 boreholes operate in the area today. The yields of these boreholes have been reduced from an average of 10 m3/h in 1980 to 1–2 m3/h in 2000. Boreholes with yields of 2–3 m3/day are still in operation. In effect, the farmers are rapidly and inexorably drying out the aquifer. The average annual extraction during the past 10 years is estimated to be around 1214 hm^3.

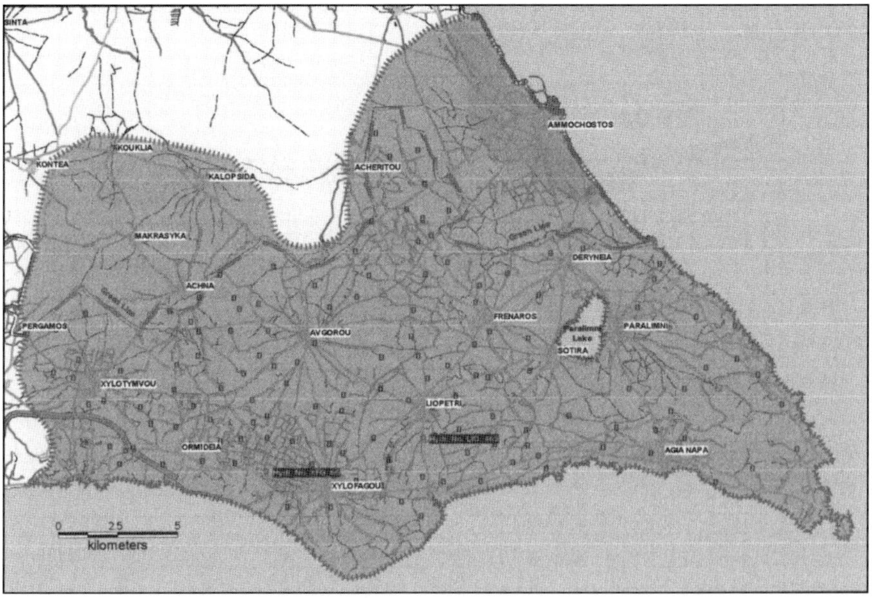

Fig. 3.22 The general area and the Kokkinochoria aquifer

Table 3.31 Recharge

Southern conveyor project comp-leted in 1987	Rainfall	Subsurface Inflow	Sea intrusion	Return from irrigation	Return imported/ Diversions	Total	Remarks (aquifer area 172 sq. km)
1963–1978 (SCP study – Iacovides)	8.2	1.1	2.9	4.7		16.9	Average rainfall 330 mm
At present (FAO study – Georgiou)	8	0.1	5.5	0.5	1.6	15.7	Average rainfall 300 mm

Table 3.32 Outflow

	Abstraction for irrigation and domestic	Sea Outflow	Total	Balance
1963–1978 (SCP study – Iacovides)	27.1	0.4	27.5	–10.6
At present (FAO study – Georgiou)	14.0	1.5	15.5	+ 0.2 *5

[5] The annual balance would be $-5.3\,hm^3$ if sea intrusion is considered. The recommended annnual pumping from this aquifer is only $8\,hm^3$.

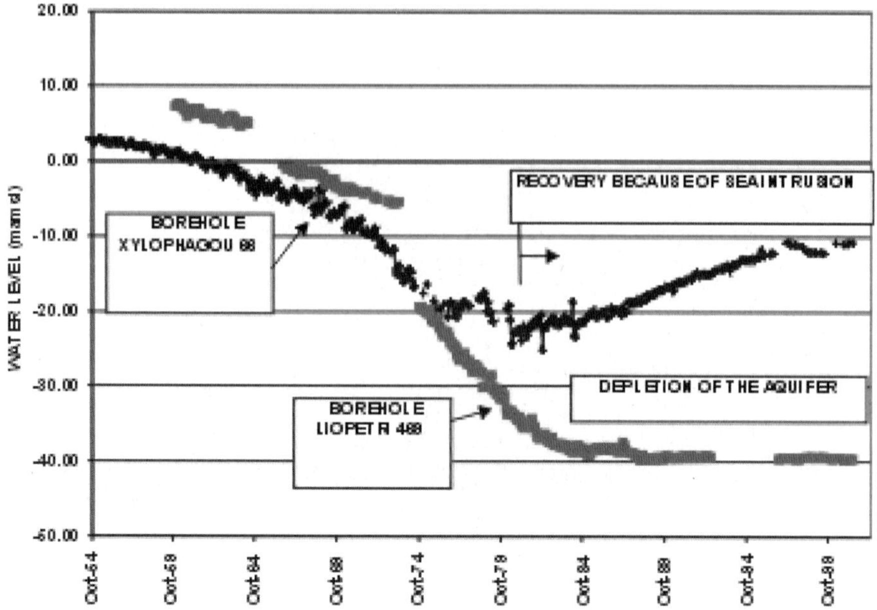

Fig. 3.23 Hydrograph of boreholes xylophagou 66 and liopetri 469 (Elev. 52.61 and 30.74m amsl)

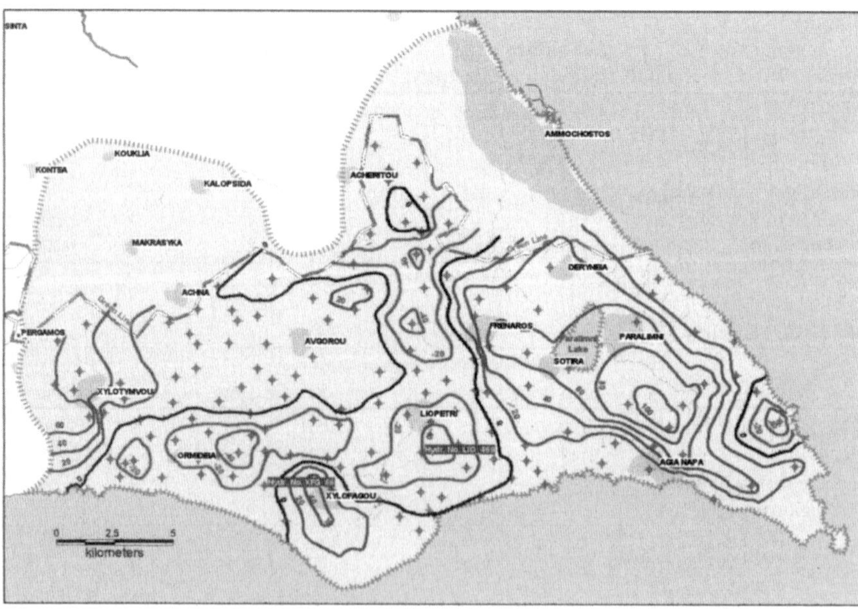

Fig. 3.24 Kokkinochoria aquifer - Water level (m amsl) contour map September 2000

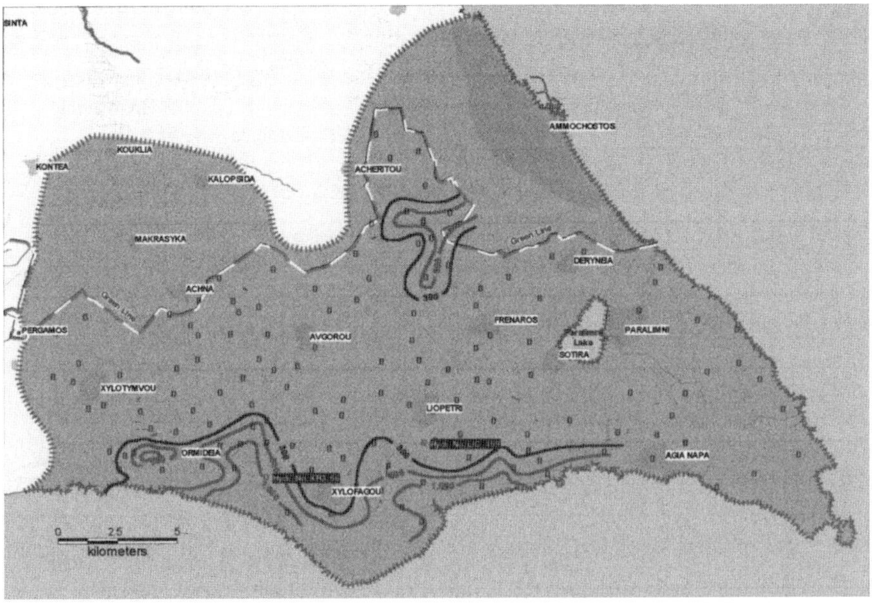

Fig. 3.25 Kokkinochoria aquifer isochloride contours (ppm) for June 1994 (after Georgiou)

Table 3.33 Kokkinochoria area matrix of circumstances

Natural conditions and infrastructure	**Regional context**	Climate type	Csa-Med/nean
		Aridity* Index	Semi-arid (0.268)
		Permanent population	30000
	Water availability	Total water resources /availability**	$30\,hm^3$**
		Trans-boundary water	–
	Water quality	Quality of surface water	Very good
		Quality of groundwater	fair
		Quality of coastal water	Good
	Water supply	Percentage of supply coming from:	
		Groundwater	30%
		Surface water	–
		Desalination, recycling	13%
		Importing	57%
		Network coverage:	
		Domestic	100%
		Irrigation	>85%
		Sewerage	Apx. 70%
	Water use	Water consumption by category:	
		Domestic	6%
		Tourism	10%
		Irrigation	84%
		Industrial and energy production	
		Resources to population index***	$1000\,hm^3/c$

(continued)

Table 3.33 (continued)

Economic and social issues	**Water demand**	Water demand trends	Increasing
		Consumption index	100%Increasing
		Exploitation index***	300%Increasing
	Pricing system	Average household budget for domestic water****	€ 99.2/yr
		Average household budget for agricultural water	€0.11/m³ depends on land
		Average household income****	€24207urban €18488 rural
		Cost recovery	Dom €0.58/m³ Irr €0.11
		Price elasticity	Very small
	Social capacity building	Public participation in decisions	Fair
		Public education on water conservation issues	Fair
Decision-making process	**Water resources management**	Water ownership	State – (partly private)
		Decision-making level regarding:	
		Water supply for each sector	National
		Water resources allocation for each Sector	National
	Water policy	Local economy basis	Agri/tertiary
		Development priorities	Agri/tourism

*Aridity = ((350 + 318 + 330)/3)/ ((4.6 + 4.4 + 4.6)/3)×365×.0.75) 1961–1990

**Includes import by SCP (17 hm^3), local groundwater (9 hm^3) and Desalination (4 hm^3)

***Exploitation index = 300% since 30 hm^3 are used against 10 hm^3 locally available

****Family Budget Survey 1996/97 Statistical Service of Republic of Cyprus

3.12 Canary Islands, Spain

Due to their geographic location, close to the Tropic of Cancer, the Canary Islands are under the influence of the trade winds, originating from the circulation of air masses around the anticyclone of Azores. Air-layering caused by trade-winds generates a characteristic layer of stratocumulus clouds on the northern coast of the higher islands, which occur between 500 m and 1,500 m. Humidity condensation in these areas entails a complementary water contribution, saving the western Canary Islands (higher than 500 m) from extreme aridity conditions.

In spite of that, the Canaries are poor in freshwater resources. The extent of their own freshwater availability, (177 m^3 per inhabitant per year), place them last in the classification by hydrographical river basins in Spain. In fact, this number is very far from the national average of 1,389 m^3/persons/year. With a population around 1.5 million inhabitants, the islands host every year more than 10 million tourists whose average daily water consumption is of 350 l/persons/day (Insular Hydrologic Plans). This increasing difference between resource availability and consumption characterizes the Canary archipelago while rainfall's main features in the region are scarcity (an average of 310 l/m^2/year) and irregularity, both in time and space.

Topographic difficulties and permeability of the existing geologic materials lead to the exploitation of a minimum share only of the surface water resources. It is explanatory enough that the volume of water retained by the some 100 dams built to this end (41 hm3/year), only reaches the 33% of their total capacity.

A relevant feature of underground water management is the fact that they are private property, a singularity in Spain. This market is subject to and regulated by the Canary Islands' Water Law (12/1990).

Another important feature that especially affects the Eastern islands is the progressive dependence on desalinated water, which is significantly increasing every year. An extreme case is the island of Lanzarote, where 97% of the water supply originates from desalination (the maximum security forecast of the system is 5.4 days). The progressive energy consumption of water desalination in the Canaries is demonstrated from the fact that in the year 2000 almost 15% of available electricity power was directed to supply existing desalination plants.

Agricultural consumption is a priority on islands like La Palma and El Hierro, reaching 80% of the total consumption. Urban and tourist consumption have a significant role on the main islands of Tenerife and Gran Canaria, where the majority of the population is concentrated. In the minor islands of the archipelago with a strong tourist penetration, tourist water consumption is progressively approaching the urban one (Lanzarote and Fuerteventura).

One of the most distinctive features of agriculture water consumption is related to the generalized presence of intensive crops characterized by a high demand. Banana plantations – representative crop and main consumer of water in the Canary archipelago – are characterized by water demands around 11,350 and 14,850 m^3/ha/year. These crops receive subventions in the framework of the European Common Agricultural Policy. In the same time, they are important producers of landscape that, similarly to other productions, have progressively reached a crisis point. This point was reached due to tourist and urban water demand conflicts, which resulted in a rise in water price, given the private character of the canary water market.

The unforeseeable population growth of the last years causes strong uncertainty as to water resource planning. In five years only the foreign population growth doubled the natural growth rate. A similar tendency is detected in the tourist sector, where the tourist lodging capacity was practically doubled in the period 1998–2001.

Regarding sewerage, we also find important deficits, especially among the dense scattered settlements of the islands, which directly influence underground water due to contamination of aquifers. Only the two main islands rely on acceptable grids, although they also have significant deficiencies in specific settlements. The remaining ones either have serious deficiencies regarding sewers or produce large quantities of surface or underground wastewater.

With regard to water treatment, serious competency conflicts have also been detected in tuning and maintenance of the treatment systems that, at present, have a very low operational rate (close to 30%). Scattered treatment plants are a distinctive feature of the Canary situation. Price policy for treated water, which is public-owned, is also characterized by variability and inconsistency. In Tenerife, for instance, the price of water treated to a third stage is €0.36 /m^3 and that of water treated to just a second stage is €0.31 /m^3. In the same time, in Gran Canaria prices are around €0.12–€0.15 /m^3, clearly below cost.

Water quality for urban supply followed a descending curve in the last years. According to following-up that was carried out by the different hydrological plans, negative effects on the quality of underground water were detected. It should be also noted that those derived from hydrogeologic situations present specific aquifers characterized by a high fluor content.

Public management, especially on a local level, faces serious difficulties that prevent sufficient and efficient implementation. Difficulties have to do, on one side, with budget origin and destination, without forgetting financial cost recovery. On the other side they are related to the political cost of transferring the cost to users, to the progressively rising prices due to the increased number of services and, finally, to the population to be served, that is characterized by a very high growth rate or by depopulation. All the above entail an influence on scale economies or diseconomies.

The studies SPA-15, Canarias Agua 2000, Mac 21, advances of several Insular Hydrological Plans and the Canary Islands Hydrological Plan constitute the basic list of water management actions and plans carried out in the Canary Islands during the last 25 years.

Within this context, the strategy of the Canary Islands Hydrological Plan is based on the following principles:

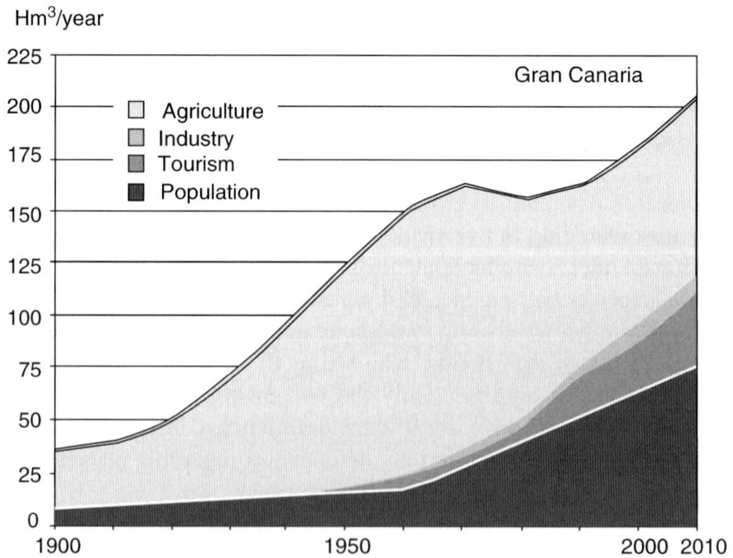

Fig. 3.26 Evolution of water demand (perspective – year 2000)
Source: Canary Island Water Centre

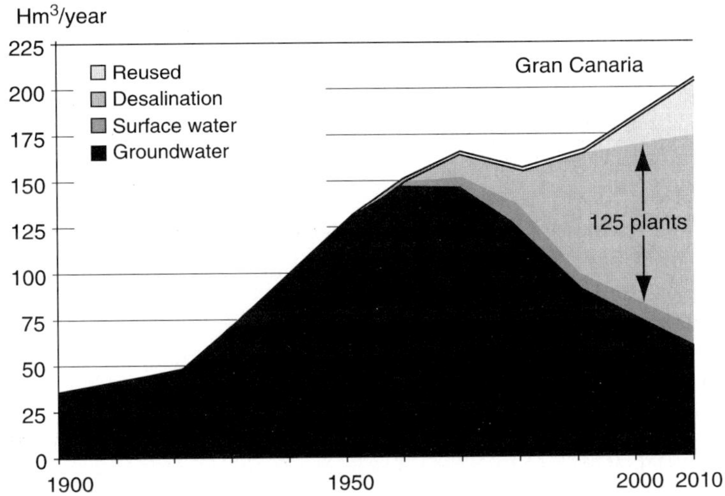

Fig. 3.27 Evolution of water production (year 2000 perspective)
Source: Canary Island Water Centre

Table 3.34 Water balance. (From Advance of island hydrological plans. 1. Island Plan of Fuerteventura s.d.– without data available.)

	Fuert-even (1)		La Gomera		Gran Canaria		El Hierro		Lanza-Rote		La Palma		Tene-rife	
Concept/Island	hm³	%	hm³	%	hm³	%	hm³	%	hm³	%	hm³	%	hm³	%
Precipitation	16	100	140	100	466	100	95.3	100	127	100	518	100	865	100
Evapotrans-piration	s.d.	–	69	49.3	304	65	69	72.4	122.2	96	238	46	606	70
Surface water	4	25	11	7.8	75	16	0.3	0.3	1.3	1	15	3	20	2
Infiltration	12	75	60	42.9	87	19	26	27.3	3.3	3	265	51	239	28

Table 3.35 Water balance

	Fuert-Event		La Gomera		Gran Canaria		El Hierro		Lanza-Rote		La Palma		Tenerife	
Production	hm³	%	hm³	%	hm³	%	hm³	%	hm³	%	hm³	%	hm³	%
Surface water Small dams	2.6	21.3	3.4	24.3	11	8.5	–	–	0,07	0.7	5	7	1	0.5
Groundwater	5.3	43.5	10.6	75.7	98	75.4	1.45	100	0.2	2.3	68	93	211	99.5
Desalination	4.3	35.2	0	–	21	16.1	–	–	9.6	97	0	–	0	–
Re-use	–	–	0	–	0	–	–	–	s.d.	–	0	–	s.d.	–
TOTAL	12.2	100	14	100	130	100	1.4	100	9.9	100	73	100	212	100

Sources: PPHH of La Palma, La Gomera and El Hierro. Hydrological Plans of Tenerife and Lanzarote;" Las Aguas del 2000" - and PIO Fuerteventura. s.d.- without data available

Table 3.36 Water consumption by category. (From PHH La Palma, La Gomera, El Hierro, Tenerife and Lanzarote; "Las Aguas del 2000".)

	Fuert-Event.		La Gomera		Gran Canaria		El Hierro		Lanza-Rote		La Palma		Tenerife	
Consumption	hm³	%	hm³	%	hm³	%	hm³	%	hm³	%	hm³	%	hm³	%
Irrigation - Agriculture	8.4	61.8	6.1	43.3	75	58	1.2	85.7	0.3	6	58	79.5	109.2	52.7
Domestic and Services	2.7	19.8	6	42.6	38	29	0.2	14.3	2.4	52	6	8.2	62.7	30.2
Tourism	2.5	18.4	–	–	15	11	–	–	1.4	31	–	–	14.1	6.8
Industrial	–	–	2	14.1	2	2	0	–	0.5	11	2	2.8	5.3	2.6
Resources non-used	–	–	–	–	–	–	–	–	–	–	6.9	9.5	4.5	2.2
Distribution losses	–	–	–	–	–	–	–	–	–	–	–	–	11.5	5.5
Total	13.6	100	14.1	100	130	100	1.4	100	4.6	100	72.9	100	207.3	100

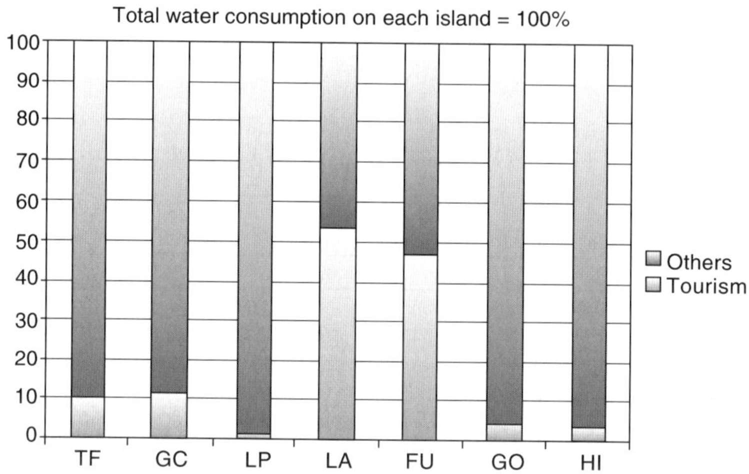

Fig. 3.28 Percentage water consumption of the tourist sector on each island
Source: Canary Island Water Centre

- To promote a sustainable use of water resources on the basis of a medium–large term planning
- To protect water ecosystems as an essential principle for a sustainable development
- To guarantee a qualitatively and quantitatively appropriate water supply to achieve a sustainable development
- To achieve the economic efficiency of water offer and use compatibly with social and environmental dimensions
- Congruence between economic and environmental criteria and the design of an integrated management system, with a prudent use of regulatory and market processes

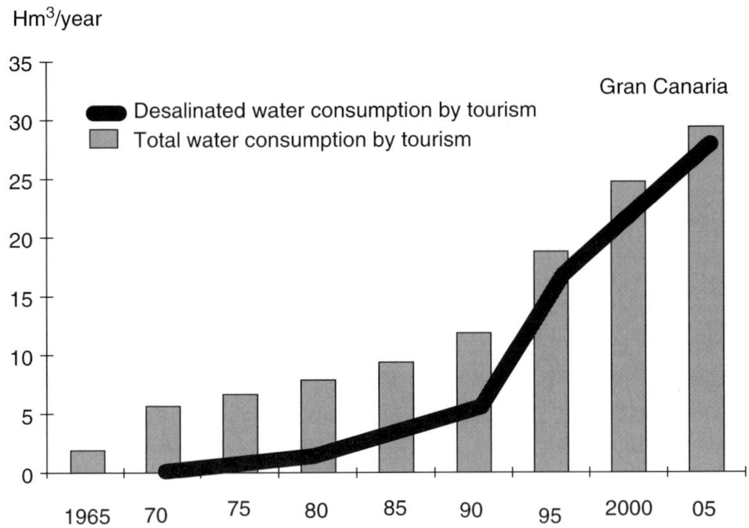

Fig. 3.29 Tourism water supply and desalination
Source: Canary Island Water Centre

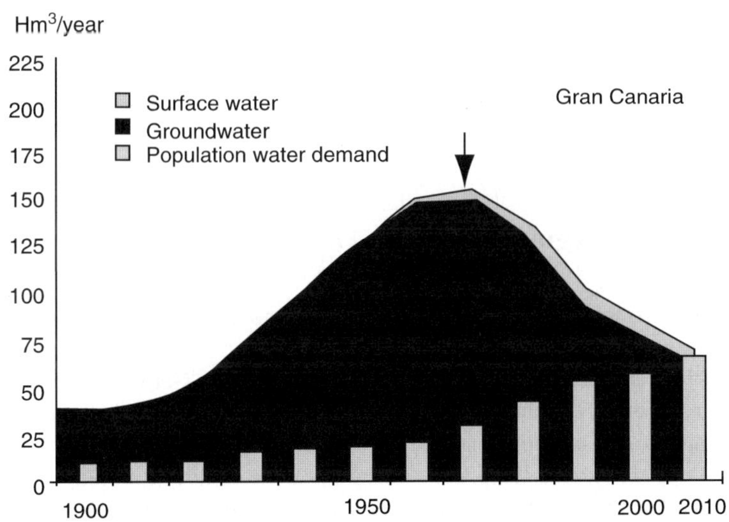

Fig. 3.30 Evolution of water production and population demand (perspective). The case of Gran Canaria
Source: Canary Island Water Centre

- To advance in setting up innovatory and realistic policies on endowment and prices
- To these criteria some considerations of the Infrastructure Director Plan within the section regarding water resources

Table 3.37 Growth in tourist accommodation, 1986–1996, Canary islands. (From White Paper on Canary Island Tourism 1998.)

Year	Tourists	Rooms
1986	4,169,050	201,493
1987	5,068,242	251,067
1988	5,416,652	308,177
1989	5,352,205	343,559
1990	5,459,473	364,269
1991	6,136,990	375,995
1992	6,327,112	337,482
1993	7,551,065	337,975
1994	9,256,817	330,614
1995	9,693,086	324,124
1996	9,804,540	328,254

- To improve knowledge about natural resources, setting up an automatic control network within the whole region that allows the following-up of comparable data and the establishment of a sound basis to achieve and maintain a sustainable use of the public water domain
- To protect quality and guarantee renovation of the different sources of production
- To optimize the implementation of systems for non-conventional resource production
- To intervene in sewerage and supply infrastructures

3.13 Doñana Region, Spain

Doñana and its surroundings constitute a natural space featuring the widest variety of pressures regarding the use and allocation of water resources. As a territory in which the most important European wetlands coexist, the National Park of Doñana includes areas of rice fields, intensive crops and a considerable tourist activity, mainly concentrated on the coastline.

Doñana could be considered to be an excellent laboratory for studying the management of water resources, a place where all the preservation and development strategies applied during the last decades share the difficulties of managing water resources.

Regarding the policies of preservation and management of water resources, plans have entirely focused on the National Park of Doñana. With an area of more than 50,000 ha, Doñana is one of the world's most emblematic coastal wetlands. Apart from being a Ramsar site and a Special Protection Area for birds, the National Park of Doñana was declared a Biosphere Reserve in 1980, and was inscribed on the World Heritage List in 1994. The Biosphere Reserve includes a

Table 3.38 Canary islands matrix of circumstances

Natural conditions and infrastructure	**Regional context**	Climate type	Oceanic
		Aridity Index	0.2 < AI < 0.6 coastal and oriental islands
		Permanent population	1,781,366
	Water availability	Total water resources /availability (hm3)	
		Groundwater	702
		Surface water	78
	Water quality	Quality of surface water	Good
		Quality of groundwater	Average
		Quality of coastal water	Poor
	Water supply	Percentage of supply coming from:	
		Groundwater	87%
		Surface water	5%
		Desalination	8%
		Network coverage:	
		Domestic	60%
		Irrigation	85%
		Sewerage	60%
	Water use	Water consumption by category:	
		Agriculture	58%
		Domestic and services	27%
		Tourism (only accommodation)	7%
		Industrial	3%
		Non-used resources	2.5%
		Losses (internal network)	2.5%
		Resources to population index	438
Economic and social system	**Water demand**	Water demand trends	Variable – Increasing
		Consumption index	53%
		Exploitation index	58%
	Pricing system	Average household budget for domestic water (pa)	356 € (Average price 1.55 m^3)
		Average household budget for agricultural Water	Variable
		Average household income	16800 €
		Cost recovery	Average
		Price elasticity	Average
	Social capacity building	Public participation in decisions	Poor
		Public education on water conservation issues	Poor
Decision-making process	**Water resources management**	Water ownership	
		Groundwater	Mostly private
		Surface water	Public and private
		Decision-making level regarding:	
		Water supply for each sector	Regional – Local
		Water resources allocation for each sector	Regional – Island
	Water policy	Local economy basis	Tourism
		Development priorities	Tourism

buffer zone of 26,000 ha, summing to a total of 77,260 ha. Doñana belongs to the small group of coastal wetlands within the three categories, together with San San-Pond Sak (Panama), Palawan (Philippines), Danube Delta (Romania-Ukraine), Ichkeul (Tunisia) and Everglades (USA).

Around this sanctuary of the European biodiversity lies the Natural Park of Doñana and its Surroundings, located in the municipalities of Almonte, Hinojos, Lucena del Puerto, Moguer and Palos de la Frontera (province of Huelva), Sanlúcar de Barrameda (province of Cádiz), Puebla del Río, Aznalcázar, Villafranco del Guadalquivir and Villamanrique de la Condesa (province of Seville). This extended list is representative to the administrative and territorial complexity of the area.

The territory occupied by Doñana's basins, which also includes the National and Natural Parks, holds over 180,000 permanent inhabitants. The figures indicate a considerable increment compared to the 128,000 inhabitants registered in 1981. More than 60% of the employment is concentrated on the agricultural sector, and another 25% is devoted to the service sector, which is mainly focused on tourism.

The agricultural development in the area arrives at a later stage due to its hard conditions: the 19thcentury witnessed a series of failed efforts oriented to drying the salt marsh. By the end of the 1920s, the area devotes itself to massive rice crops, which nowadays occupy over 35,000 ha, thus becoming a factor of pressure for the National Park.

After this episode, in the 1970s, the FAO generates a report that results in the creation of a Plan for Agricultural Development in Almonte-Marismas (decree 1194/71), driven by a development-oriented mentality that resolves to declare it an Area of National Interest. This is the consolidation of 45,960 ha of crops; 30,000 of which correspond to irrigated land. This strategy is based on recognizing the existence of an important water table in the area. Nowadays, the useful surface for irrigation sums up to approximately 14,000 ha.

Regarding the agricultural exploitation, we must highlight de importance of the strawberry trees, which occupy some 2,500 ha, and constitutes a very concentrated source of employment. The exploitation of groundwater does not directly affect the water supply of the National Park, although it does affect the quality of underground waters, which sometimes feature nitrate concentrations of more than 50 mg/l.

The tourist activity, mainly concentrated on the area of Malascañas, located at the border of the National Park, is also a factor of pressure for water resources, especially during times of drought. Matalascañas offers a tourism capacity of 63,233 people, with a high level of concentration during the high season.

All these episodes resulted in an alteration of the water regimes, followed by a serious overexploitation of groundwater and manipulation of superficial water systems, which have seriously endangered the preservation of the National Park of Doñana.

This has lead to a progressive recognition of the fact that the preservation of the National Park is not only an obligation brought about by the need to preserve this important natural sanctuary, but also by the fact that Doñana is of patrimonial value which cannot be dissociated from the future economy of the area. This concern has resulted in the implementation of several strategies oriented to the sustainable management of water resources. In this sense, we must highlight the International

Experts Commission's Report about the Development of Strategies for the Sustainable Development of Doñana in 1992. This report has inspired many of the principles for the alternative management of water resources during the last years.

In 1998, though, Doñana faced one of its worst moments due to the breaking of a pyrite pond belonging to a mining exploitation, which caused the flooding of more than 2,600 ha with high metal content muds. Although the muds did not reach the park itself, this accident caused red alert within all administrations and the whole society. After an impressive deployment of technical and human resources, the muds could be removed avoiding an ecological catastrophe with unforeseeable consequences.

What at the beginning appeared to be one more regrettable accident due to lack of planning and foresight in natural areas management turned to be the start of one of the most important wetland regeneration initiatives ever carried out in the whole planet. In reply to this situation, the big water regeneration programme named "Doñana 2005" started, supported by the Spanish Ministry of Environment, whose immediate environmental actions were funded with some 140 million €. It is a project whose objectives are much more ambitious than providing the mere solution for the problems caused by the accident. It is also complemented by another important action called "the Green corridor of Doñana", supported by the "Junta de Andalucía" that will be carried out within the buffer zone.

3.13.1 Hydrological Characteristics

The area is divided into two domains. The first one is the salt marsh which is a very plain area that combines periods of flood and drought. Its main sources of water are the rivers and tributaries and, in a smaller proportion, some few emergencies of underground water running through pipes. The rest of the territory is basically made up of sand. This is the area where water precipitations overload the water table (called water Table 3.27). It holds most of the water demanding activities.

On the overall system, the role and the alteration of underground waters is one of the fundamental problems for the management of this resource in the area. As in many other groundwater cases, overload is one the factors where estimations are subject to error. The figures range from 50 to over 200 mm/year.

3.13.2 The Challenges

The conflict between preservation and a balanced leverage of water resources in Doñana materializes with solving and recognition of the following aspects:

- The overexploitation of groundwater is seriously affecting natural areas of vital importance. The effects of overexploiting the underground waters in the ecosystems seem to be put off with the years. Nowadays, a great portion of the water table under the salt marsh has fallen from a 1-m level over the ground to

Table 3.39 Donana matrix of circumstances

Natural conditions and infrastructure	**Regional context**	Climate Type	Mediterranean
		Aridity Index	0.4 < AI < 0.65
		Permanent population	180,000
	Water availability	Total water resources /Availability (hm3)	Min: 155 hm^3/year
		Groundwater	Max: 425 hm^3/year
		Surface water	Min: 32 hm^3/year
			Max: 78 hm^3/year
	Water quality	Quality of surface water	Average
		Quality of groundwater	Average
		Quality of coastal water	Average
	Water supply	Percentage of supply coming from:	
		Groundwater	97%
		Surface water	3%
		Desalination	0%
		Network coverage:	
		Domestic	95%
		Irrigation	95%
		Sewerage	60%
Economic and social system	**Water use**	Water consumption by category:	
		Agriculture	84%
		Domestic and services	4%
		Tourism (only accommodation)	8%
		Industrial	1%
		Non-used resources	3%
		Losses (internal network)	30%
		Resources to population index	
	Water demand	Water demand trends	Variable – Increasing
		Consumption index	53%
		Exploitation index	Max: 49%
	Pricing system	Average household budget for domestic water (pa)	€50
		Average household budget for agricultural water	€8,114
		Average household income	€7.535
		Cost recovery	Average
		Price elasticity	Fix
	Social capacity building	Public participation in decisions	High
		Public education on water conservation issues	Average
Decision-Making Process	**Water resources management**	Water ownership	
		Groundwater	Public and private
		Surface water	Public
		Decision-making level (municipal, regional, national) regarding:	
		Water supply for each sector	Regional – Local
		Water resources allocation for each sector	Basin
	Water policy	Local economy basis	Agriculture
			Tourism
		Development priorities	Intensive agriculture
			Tourism

a 2-m fall under its own level. Overexploitation is starting to allow the entrance of the salty waters contained in the sediments of the salt marsh, with a considerable impact on the water quality.

- The massive usage of fertilizers in the main agricultural activities has a devastating effect on the water quality.
- The organic contribution due to domestic tributaries also adds to the problem, since the network of cleansing stations is still to be completed.
- The agricultural and industrial residues, especially vegetable water deriving from olive manipulation, result in scattered episodes of contamination in large brooks.

The original water system of the salt marsh is deeply altered. For many years, balancing the complex system of the salt marsh was attempted by a series of corrective actions. A considerable part of the Doñana Programme 2005 is oriented to regenerating the hydrological systems for the basic functions of the salt marsh and making it compatible with human needs.

3.14 Sado Region, Portugal

Sado's river basin has an area of 8,295 km^2 and a population of 292,960 inhabitants (1998). Its population density of about 35 inhabitants per km^2 is a low value when compared to Portugal Continental territory value of 110 inhabitants per km^2. The region is mostly plain, except to some low mountains, with an overall average altitude of 127 m. In fact, altitudes range from 50 m to 200 m in most of the area, with a maximum basin altitude of 501 m. Most of the area presents tertiary and quaternary deposits with formations mainly composed by limestone and sedimentary rocks.

The climate is Mediterranean temperate, with rainy winters and dry summers. The average temperature is of 16 °C, and in the summer peak months (July and August), it varies from 19 °C in the coastal areas to 24 °C in the interior. In the coldest month (January) it varies from 9 °C in the interior to 12 °C in the coastal areas. The average annual sunshine duration is about 2,900 h. The average annual precipitation is 622 mm, ranging from less than 600 mm in the coastal areas to more than 900 mm on the mountains. About 78% of the precipitation is concentrated in the dry semester (between October and March) and occurs 75–100 days per year in the coastal areas and 50–75 days per year in the rest of the basin. As to potential evapotranspiration, its yearly average is 1145 mm increasing in the dry semester. Fig. 3.31 presents the mean monthly precipitation and potential evapotranspiration in Sado's basin. The total runoff is 972 hm^3/year.

Sado has a storage capacity of 771 hm^3 which makes it the Portuguese river basin with the biggest storage capacity when compared to mean annual flow, reflecting irrigation availability needs. The overall availability is currently of 1,714 hm^3/year, in average, consisting of 918 hm^3 of surface water (716 hm^3/year in dry years) and 796 hm^3 of exploitable groundwater. However there will be a big increase in the availability of surface water, due to the foreseen inter-basin water transfer of about

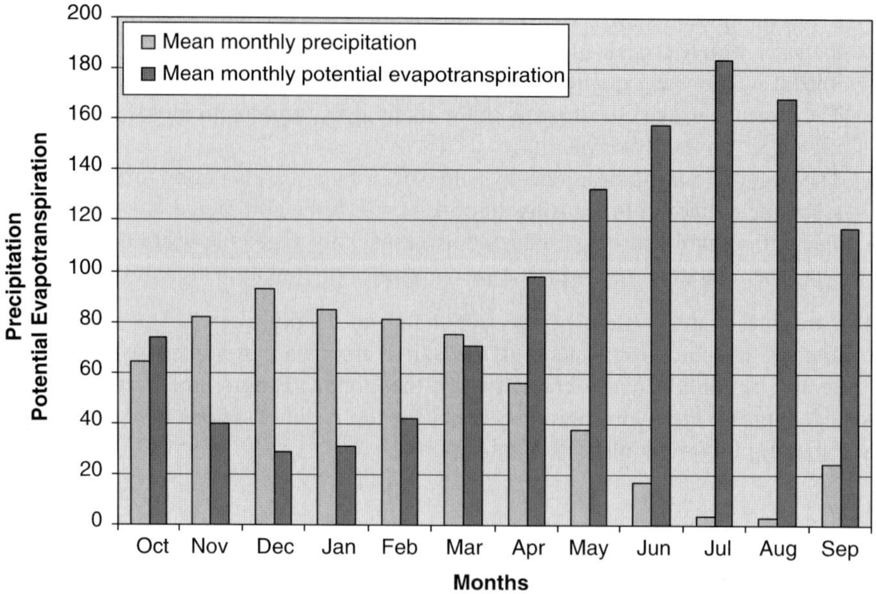

Fig. 3.31 Mean monthly precipitation and potential evapotranspiration in Sado's basin

450 hm^3/year from Guadiana's basin, when Alqueva new multi-purpose hydraulic plant (still in construction) will start operating.

Surface water is considered inadequate for the various uses, according to the national legislation, due to poor quality, with concentrations exceeding the recommended values. In terms of groundwater, in the monitored aquifers, water quality is good. As to coastal waters the quality is also good, with the exception of one or two polluted spots.

Industry, animal husbandry and non-point source loading from agriculture are responsible for the majority of the pollution loads verified in Sado, which for 1998 have been estimated as:

- BOD5 = 22,461 ton/year
- TSS = 45,281 ton/year
- COD = 42,807 ton/year
- Total nitrogen = 3,926 ton/year
- Total phosphorus = 2,505 ton/year

The share of population served with water supply is currently of 97%, higher than the value corresponding to wastewater drainage (87%), with only 56% benefiting of treatment facilities. Overall losses in urban water supply are currently high (average of 20%), and there is a low overall efficiency in agriculture water use (about 60%). The total annual water consumption is 1,195 hm^3 (600 hm^3 of return flow), distributed as follows: 672 hm^3 of water are used in energy production to the cooling in thermo-electric power plants, 441 hm^3 in agriculture, 58 hm^3 in industry (mainly Sines' industry on the coast), and about 24 hm^3 in domestic uses, with the water uses in tourism less than 1 hm^3. The total water consumption in the area represents 12.03% of the total water use in Portugal Continental.

The average household income in 2000 was €13,562 /year, with only 0.75% of this value allocated to domestic water supply, which indicates a low water pricing (€0.57 /m^3) and a low urban water sector cost recovery (37%). This situation is much aggravated in agricultural sector, with prices very low (€0.06 /m^3) and strongly subsidized.

There is no inter-municipal primary urban water supply system covering the basin. The (secondary) water supply distribution networks are almost 100% (except one system, partly owned by Águas de Portugal group) of full municipal responsibility. The situation is similar with respect to wastewater drainage and treatment systems. Thus the pricing of water is mostly a political issue and not currently aiming at cost recovery.

Table 3.40 Sado matrix of circumstances

Natural conditions and infrastructure	**Regional context**	Climate type	Cs: Mediterranean temperate
		Aridity Index	AI = 0.54 Dry Sub-humid
		Permanent population	292,960
	Water availability	Total water resources/ Availability (hm^3)	1768 /1714
		Trans-boundary water	No
		Inter-basin water transfer	Yes (−2 hm^3 (*) −10 hm^3 (**))
	Water quality	Quality of surface water	Poor
		Quality of groundwater	Good
		Quality of coastal water	Good
	Water supply	Percentage of supply coming from:	
		Groundwater	16%
		Surface water	84%
		Desalination, Recycling	–
		Importing	–
		Network coverage:	
		Domestic	97%
		Irrigation	72%
		Sewerage	87%
	Water use	Water consumption by category: (hm^3)	
		Domestic	24.3
		Tourism	0.6
		Irrigation	441
		Industrial and energy production	730.1
		Resources to population index	6035

(continued)

Table 3.40 (continued)

Economic-and Social System	**Water demand**	Water demand trends	Increasing
		Consumption index	68%
		Exploitation index	70%
	Pricing system	Average household budget for domestic water	0.75%
		Average household budget for agricultural water	€0,06/m^3
		Average household income	€13562/year
		Cost recovery	Low (37%)
		Price elasticity	Very small
	Social capacity building	Public participation in decisions	Poor
		Public education on water conservation issues	Poor
Decision-making process	**Water resources management**	Water ownership	Public (partly private)
		Decision-making level regarding:	
		Water supply for each sector	National/Municipal
		Water resources allocation for each sector	National
	Water policy	Local economy basis	Agriculture and industry
		Development priorities	Agriculture

*Transferred to Guadiana's basin

**Transferred to Rib. Costa Alentejo basin

3.15 Guadiana Region, Portugal

Guadiana's[4] river basin covers an area of 11,601 km^2 and has a population of 182,580 inhabitants (1998), with a population density of about 16 inhabitants per km^2, almost the lowest value of all Portuguese river basins. The average altitude is 237 m and most of the region altitudes range from 100 m to 400 m, with a southern mountain chain (making the division between Alentejo and Algarve) where the maximum altitude occurs (1,027 m). The slopes are mainly of 0%–5% with 5%–30% in the mountains. Most of the area presents formations mainly composed by metamorphic, eruptive and sedimentary rocks, with 2/3 of the basin composed by schistones.

The climate is temperate, with rainy winters and hot and dry summers. The average temperature is 16 °C, and in the summer peak months (July and August) it varies from 23 to 26 °C. In the coldest month (January) it varies from 8 °C in the north of the basin to 11 °C in the (south) coastal areas. In this river basin temperature reaches maximals of 41 to 44 °C. The average annual sunshine duration is 2,829 h, with a maximum for July (370 h) and minimum for December (147 h). The average annual precipitation is 568 mm, spatially ranging from a minimum of

[4] Northeastern Parnitha not included as it discharges in the Water Region of Eastern Sterea Ellada.

350 mm to little more than 1,000 mm. Precipitation occurs 5080 days per year and, in volume, more than 80% of it is concentrated in the dry semester (between October and March). As to potential evapotranspiration, the averaged yearly value is 1,242 mm and it increases in the dry semester. Fig. 3.32 presents the mean monthly precipitation and potential evapotranspiration in Portuguese Guadiana's basin territory. The total runoff due to that part of the basin is 1,887 hm^3/year, whereas in Spain the annual mean flow is 5,470 hm^3/year.

Portuguese Guadiana's water storage capacity is of 460 hm^3, but this figure will be highly increased due to Alqueva new multi-purpose hydraulic plant (still in construction), which will account for a (useful) storage capacity of 3,150 hm^3. Although currently a 30 hm^3 inter-basin water transfer from this basin to Algarve occurs, Alqueva's storage capacity will enable a big inter-basin water transfer of about 700 hm^3, mainly to Algarve (for irrigation and for public water supply, in support of tourism water needs) and also to Sado's basin (for domestic and industrial water supply).

In terms of water availability, in average, it is currently of 3,585 hm^3/year, consisting of 3,156 hm^3 of surface water (1,476 hm^3/year in dry years) and 429 hm^3 of exploitable groundwater. It should be emphasized that Guadiana's basin is one of the regions of Portugal that has lately been most affected by droughts, namely in the beginning of last decade (90–95), when periods with no affluences occurring from Spain. This way, the "resources to population index" was evaluated on a dual way, i.e., considering "total resources" as the natural mean flow (i) of the whole

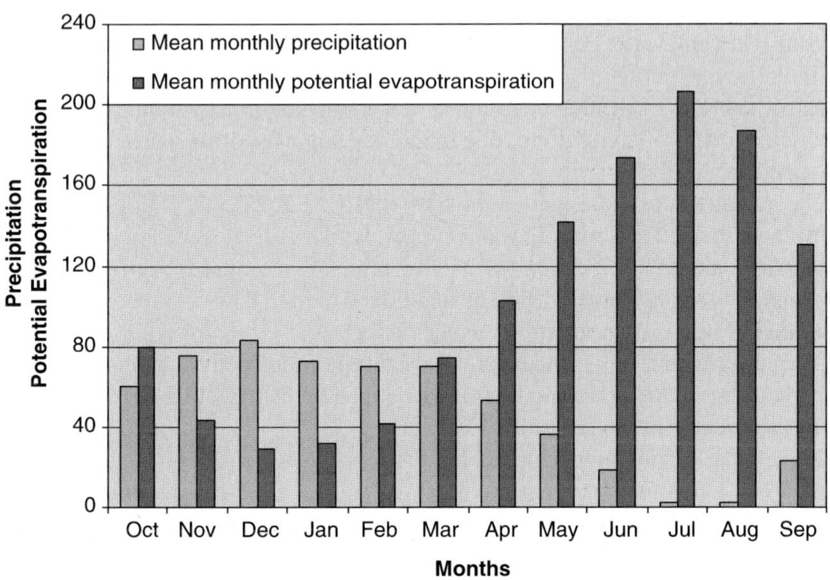

Fig. 3.32 Mean monthly precipitation and potential evapotranspiration in Guadiana's basin

basin or (ii) of the Portuguese basin area only. In a similar way, this has been reflected also on the consumption index, with values of (i) 5% and (ii) 18%, respectively. Nevertheless, although those values are currently not high, it should be stressed that even if only the referred 700 hm^3 inter-basin transfer (to Sado and Ribeiras do Algarve) is taken into account, those values would increase to values of (i) 14% and of (ii) 49%, correspondent to the two different "total resources" definitions.

Surface water is considered inadequate to the various uses, according to the national legislation, due to its poor quality, as it is the receptor of the pollution caused mostly by Spain but also by national agriculture. The same can be applied to some aquifers, with groundwater presenting concentrations of magnesium, sodium and nitrates that exceed the maximum acceptable values for drinking water. As to coastal waters, quality is good.

The main pollutant loads produced in Guadiana's basin in 1998 were estimated as:

- BOD5 = 17,389 ton/year
- TSS = 17,849 ton/year
- COD = 26,250 ton/year
- Total nitrogen = 6,425 ton/year
- Total phosphorus = 2,194 ton/year

The percentage of population served with water supply is currently of 84% and a similar percentage (83%) applies to wastewater drainage, but only 67% benefit from treatment facilities. There is a low overall efficiency in agriculture water use (about 60%). The total annual water consumption is 419 hm^3 (with about 98 hm^3 of return flows), with the following distribution: about 400 hm^3 of water are used in agriculture, 14.5 hm^3 in domestic uses, 3.3 hm^3 in industry and about 1.4 hm^3 in tourism. The total water consumption in the area represents 4.57% of the total water use in Portugal Continental territory. Again it should be stressed that water demands are expected to increase due to Alqueva new multi-purpose hydraulic plant in construction, namely on agriculture, due to the development of the currently predicted new irrigation areas.

The average household income in 2000 was €13,562 /year, with only 0.89% of this value allocated to domestic water supply, which indicates the low water pricing (€0.70 /m^3), although much higher than the price of water used for irrigation(€0.06 /m^3), which is strongly subsidized, as mentioned earlier. The cost recovery is correspondingly low (23% in the urban water sector).

Agriculture indirectly is the most important economic activity in the region, with the viticulture sector assuming high importance and contributing to the increase of the tertiary sector in the region.

There is a recent inter-municipal urban (main) water supply and wastewater drainage and treatment systems' company (Águas de Portugal group) that only "covers" the northern part of the basin. Thus the pricing of water is (still) mostly a political issue and not currently aiming at cost recovery.

Table 3.41 Guadiana matrix of circumstances

Natural conditions and infrastructure	**Regional context**	Climate type	Csa (Temperate)
		Aridity Index	AI = 0.46 Semi- Arid
		Permanent population	182,580
	Water availability	Total water resources/ Availability (hm3)	7,800 (2,300)(*)/3,585
		Trans-boundary water	5,500
		Inter-basin water transfer	Yes (2 hm^3 (**) –30 hm^3 (***)
	Water quality	Quality of surface water	Poor
		Quality of groundwater	Poor
		Quality of coastal water	Good
	Water supply	Percentage of supply coming from:	
		Groundwater	55.5%
		Surface water	56%
		Desalination, Recycling	–
		Importing	0.5%
		Network coverage:	
		Domestic	84%
		Irrigation	76%
		Sewerage	83%
Economic and social system	**Water use**	Water consumption by category: (hm^3)	
		Domestic	14.0
		Tourism	1.37
		Irrigation	400
		Industrial and energy production	3.3
		Resources to population index	42,720 (12,600)(****)
	Water demand	Water Demand trends	Increasing
		Consumption index	5.4% (18.2%)(****)
		Exploitation index	12%
	Pricing system	Average household budget for domestic water	0.89%
		Average household budget for agricultural water	0,06 €/m^3
		Average household income	13562 €/year
		Cost recovery	Low (23%)
		Price elasticity	Very small
	Social capacity building	Public participation in decisions	Poor
		Public education on water conservation issues	Poor
Decision-Making Process	**Water resources management**	Water ownership	Public (partly private)
		Decision-making level regarding:	
		Water supply for each sector	National/Municipal
		Water resources allocation for each sector	National
	Water policy	Local economy basis	Agriculture/Tertiary sector
		Development priorities	Agriculture

*Internal resources

**Imported from Sado's basin

***Transferred to Ribeiras do Algarve basin

****Considering "Total Resources" as "Internal Resources"

3.16 Ribeiras do Algarve Region, Portugal

Ribeiras do Algarve river basin covers an area of 3,837 km^2 and its population is 324,100 inhabitants (1998), with a population density of about 84 inhabitants per km^2, a value still smaller than the average for Portugal Continental territory (110 inhabitants per km^2) but the greatest among all southern (of Tejo) river basins, namely Sado and Guadiana. The region is mostly plain, with altitudes ranging from 0 to 100 m, and only a few spots above these values. Most of the area presents formations mainly composed by volcanic rocks (especially basalts).

The climate is Mediterranean temperate, characterized by rainy winters and dry summers. The average temperature is 18 °C. The average annual sunshine duration is maximal along the south coastal areas (3,180 h). The average annual precipitation is 840 mm, occurring 50–75 days per year in almost all the region. As to potential evapotranspiration its yearly average is 1,229 mm and it increases in the dry semester. Fig. 3.33 presents the mean monthly precipitation and potential evapotranspiration in Ribeiras do Algarve basin. The total runoff is 348 hm^3/year.

The overall water availability is currently, in average year, of 599 hm^3/year, consisting of 327 hm^3 of surface water (160 hm^3/year in dry years) and 272 hm^3 of exploitable ground water. Algarve's storage capacity is small (about 63 hm^3).

Surface water presents quality problems since rivers have almost no flow in dry period and receive the pollution caused by urban areas and agriculture. Dam storage reservoirs assume a high importance in water supply due to that fact, but also some

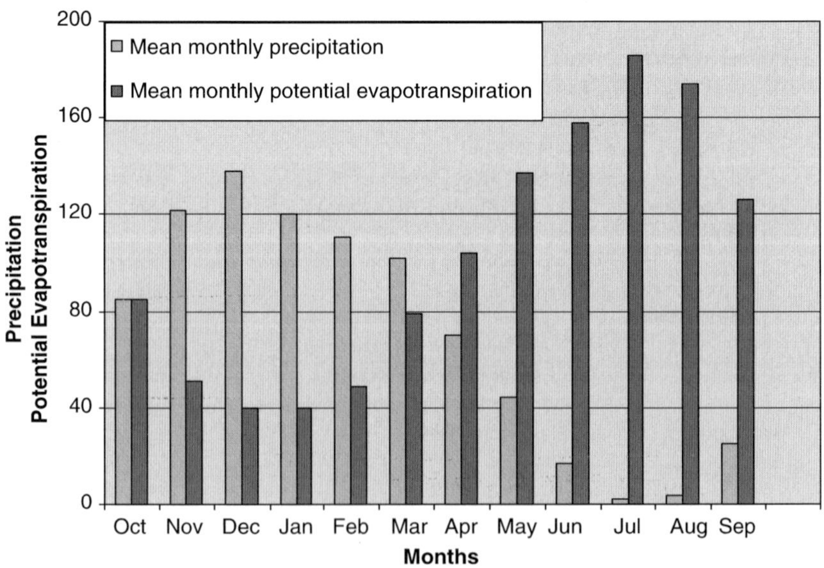

Fig. 3.33 Mean monthly precipitation and potential evapotranspiration in Ribeiras do Algarve basin

Table 3.42 Ribeiras do Algarve matrix of circumstances

Natural conditions and infrastructure	**Regional context**	Climate type	Cs: Mediterranean Temperate
		Aridity Index	AI = 0.68
		Permanent population	324,100
	Water availability	Total water resources/ Availability (hm3)	620/599
		Trans-boundary water	No
		Inter-basin water transfer	Yes (30 hm^3) (*)
	Water quality	Quality of surface water	Poor
		Quality of groundwater	Poor
		Quality of coastal water	Good
	Water supply	Percentage of supply coming from:	
		Groundwater	71.5%
		Surface water	19.7%
		Desalination, recycling	–
		Importing	8.8%
		Network coverage:	
		Domestic	82%
		Irrigation	77%
		Sewerage	73%
Economic and social system	**Water use**	Water consumption by category: (hm^3)	
		Domestic	21.8
		Tourism	10.9
		Irrigation	305
		Industrial and energy production	2.4
		Resources to population index	1912
	Water demand	Water demand trends	Increasing
		Consumption index	55%
		Exploitation index	57%
	Pricing system	Average household budget for domestic water	0.90%
		Average household budget for agricultural water	€0.07/m^3
		Average household income	€13,573/year
		Cost recovery	Low (40%)
		Price elasticity	Very small
	Social capacity building	Public participation in decisions	Poor
		Public education on water conservation issues	Poor
Decision-making process	**Water resources management**	Water ownership	Public (partly private)
		Decision making level regarding:	
		Water supply for each sector	National/Municipal
		Water resources allocation for each sector	National
	Water policy	Local economy basis	Tourism
		Development priorities	Tourism and agriculture

*Imported from Guadiana's basin

quality problems occur on it, especially in the summer. The same can be applied to groundwater with parameters like calcium, sodium, chlorides and nitrates exceeding the maximum acceptable values for drinking water and irrigation water. As to coastal waters the quality is good, with the exception of one or two polluted spots.

The main pollutant loads are mostly generated by urban wastewater, animal husbandry and non-point source loading from agriculture. The loads produced in Ribeiras do Algarve Basin in 1998 were estimated as:

- BOD5 = 11,678 ton/year
- TSS = 17,492 ton/year
- COD = 12,091 ton/year
- Total nitrogen = 2,140 ton/year
- Total phosphorus = 980 ton/year

The percentage of population served with water supply is currently of 82%, higher than the value correspondent to wastewater drainage (73%), with only 72% benefiting of treatment facilities. Urban water supply overall losses are currently high (with an average of 37%) and there is a low overall efficiency in agriculture water use (about 60%). The total annual water consumption is 340 hm^3 (about 95 hm^3 returning back to the hydric environment), distributed as follows: 305 hm^3 of water are used in agriculture, 21.8 hm^3 in domestic uses, 11 hm^3 in tourism, and in industry about 2.4 hm^3. The total water consumption in the area represents 3.74% of the total water use in Portugal Continental territory.

Water shortage occurs in the summer period, when the demands of water are higher, since this is an area that attracts a large number of tourists (currently estimated as 780,000, more than twice the permanent population) and also having increased needs of water for irrigation. Thus, a conflict is raised between the two sectors.

The average household income in 2000 was €13,573 /year, with only 0.90% of this value allocated to domestic water supply, which indicates a low water pricing (€0.68 /m^3), although much higher than for irrigation water (€0.07 /m^3), which is strongly subsidized, as mentioned earlier. The cost recovery is correspondingly low (40% in the urban sector).

There are two inter-municipal urban water supply, wastewater drainage and treatment systems covering most of the region, both under a company (Águas de Portugal group) which is the responsible for the "primary system" overall management. The secondary (domestic) water supply and wastewater drainage systems are of municipal responsibility. Thus the pricing of water is (still) a political issue and not currently aiming at cost recovery.

Chapter 4
Landscape Sensitivity, Resilience and Sustainable Watershed Management

James McGlade, Brian S. McIntosh, and Paul Jeffrey

4.1 Introduction

The conservation and future sustainability of vulnerable fluvio-coastal environments, along with the need for viable planning criteria and policy instruments for their long-term management, are some of the central issues at the heart of the contemporary environmental discourse.[1] For example, in the Mediterranean, coastal, riverine and wetland areas are subject to increasing and unprecedented changes, as a consequence of human-induced processes, such as industrial activities, commercial harbour construction, land reclamation, drainage, canal construction and growing urban encroachments (Falkenmark and Lindh, 1993; Breton, 1996; Breton et al., 1996). But perhaps the single most important threat to sustainability is to be seen in the effects of a rapidly expanding tourist sector, along with its attendant hotel and service industries and their ever-growing demands for water – something particularly acute in semi-arid regions of Spain (Breton and Sauri, 1997). What is most worrying about such a situation and one that has largely developed over the last 40 years, is that historically such developments have frequently occurred in the absence of adequate planning and environmental controls. Indeed, in many cases, land-use planning has been short-termist, and decisions have been retro-active; that is, they have been concerned with 'sticking plaster' or coping solutions, rather than the implementation of long-term adaptive management strategies. An inevitable consequence of this tradition of *ad hoc* policymaking – and particularly the encouragement of mass tourist developments – has been the dramatic increase in pollution, soil erosion, pressure on water consumption and general degradation of the environment including its cultural and natural heritage (Pearson and Sullivan, 1995;McGlade, 2001a).

Indeed, in the semi-arid areas of the Mediterranean, particularly in the Middle East and Spain, water is even more of a critical commodity because of the extreme variations in rainfall and the ever-present threat of drought. Consequently there is a constant danger of conflict in river catchments and coastal regions where water

[1] As a key concern of the European Community, these issues have been enshrined in legislation such as the Treaty on European Union (Article 130S).

P. Koundouri (ed.), *Coping with Water Deficiency*, 113–134.

supply and use is contested (Bulloch and Darwish, 1993; Smith, 1997). One of the most significant responses to this situation is to be seen in attempts to encourage the construction of integrated approaches to coastal zone and river basin management. These have stressed the need for coherent planning methods and cross-disciplinary approaches to data acquisition[2]. On the other hand, while much work has been devoted to the development of legislation and policy instruments within existing integrated coastal zone management schemes, nonetheless, they are often ineffective due to the lack of coordination between the various actors and institutions and their often conflicting world views. Moreover, the much-voiced support for cross-disciplinary cooperation is frequently not matched by practical action.

Perhaps the most significant barrier to addressing these issues stems from a lack of holistic thinking at governmental and managerial levels. In essence, this is due to a low-level understanding of the nature of complexity and the nonlinear connectivities that structure socio-natural systems; for example, solutions are often sought in large-scale decision-support systems models that generally are ill equipped to account for the levels of complexity involved, especially the array of power structures and counter-intuitive behaviours displayed by socio-political organisations, and/or the vested interests of individuals. In particular, there are frequently fundamental conflicts between those stakeholders focused on political and economic concepts of growth and others committed to approaches favouring conservation, the maintenance of biodiversity and local scale interventionist strategies. Significantly, these conflicts operate at local, regional, national and European scales and reflect fundamental differences in perception and value systems. Thus a crucial issue, for any conception of sustainable management, is the need to understand the socio-environmental driving forces of change at different spatio-temporal scales. What this means is an ability to assess the resilience of socio-natural landscapes to a variety of human and naturally induced pressures – effectively, developing an understanding of the variable sensitivities of ecological, economic and socio-cultural processes, so as to anticipate likely future outcomes and possible unforeseen development trajectories.

We address these issues by taking a critical look at the theoretical basis within which current research on socio-natural systems is undertaken. Specifically we focus on the relationships between resilience (as a manifestation of sustainability) and the notion of 'landscape sensitivity', assessing its potential usefulness as a theoretical construct that might contribute to a better understanding of watershed dynamics, in climatically marginal environments.

4.1.1 The Dimensions of Sustainability

Since it is impossible to separate landscape sensitivity issues from their wider context within the general sustainability discourse, we shall begin by examining the nature of sustainability as it impinges on our discussion. In presenting such a discussion,

[2] For example, the EC Programme Towards Sustainable Development, 1993 and the 'Proceedings of the Corfu European Summit', 1994.

we seek a rapprochement between a wide spectrum of views and value systems that define the actions of scientists, managers, politicians, farmers and urban communities. Renewable resources, sustainable economic development, employment security, conservation of the cultural landscape, these are all voiced as preferred desires – and in some cases, demands. Their coexistence, however, is problematic and as we shall see later, requires carefully negotiated solutions.

It is something of a paradox that despite the wide coverage and prominence of the theme, 'sustainability' yet remains an exceedingly ambiguous term, occupying a territory in which it appears to be 'all things to all people'. Any survey of the literature necessarily must conclude that sustainability is best described as having an 'elastic' meaning, eminently malleable and infinitely variable in usage. Thus it can be invoked to support a variety of positions depending, for example, on our valuation of natural and man-made capital (e.g. Daly, 1994; Faucheux and O'Connor, 1998).

4.1.2 Sustainability: Some Problems

Broadly speaking, and with respect to ideas of sustainable development promoted by the *World Conservation Strategy*, sustainability should meet basic human needs while maintaining the basic life support systems along with ecological diversity (IUCN/UNEP/WWF, 1980). As is well known the real popularisation of these concepts was the result of the document produced by the *World Commission on Environment and Development*, from which derives the classic definition of sustainable development as: 'development that meets the needs of the present without compromising the ability of future generations to meet their needs' (WCED, 1987: 8).

Most important, such a definition presages a shift towards putting decision-making into the hands of local communities, as opposed to national and/or supra-national bodies. Thus the idea of 'sustainable futures' such as it has any meaning, is inextricably related to a decentralisation of power so individual localities assume responsibility for the management of their resources – a point later enshrined as Local Agenda 21 at the Rio summit. Importantly, such a structure is not meant as a replacement for management at supra-regional or national scales, rather it suggests the need for more local interventionist methods in landscape planning, as an important aspect of community well-being, for the health of the ecosystem and the maintenance of biodiversity.

These are bold ideas and it needs to be said that they have so far failed to be implemented to any satisfactory degree. In fact these ideas, which are central to any restructuring of human-environment issues, have effectively been marginalised; the debate has been hijacked by the search for rigorous quantifiable criteria, to cater for a scientific agenda which needs 'answers'. For example, the growing sub-fields of ecological economics and environmental impact assessment have sought to determine appropriate economic values for the natural world; *ergo* each object of nature – be it tree, river, mountain, coastal zone, etc., has a potential dollar value which can be discounted against its exploitation or harvest. The utilitarian philosophy which underpins these approaches dominates the sustainability discourse, while issues of ethics and responsibility are secondary concerns – relegated as epiphenomena with respect to the more central quantitative concerns of predictive science and the pursuit of 'solutions' and 'answers'.

Within such a model, the worlds of agency, of communicative action, of values and intentionality are set aside from many environmental debates for they belong to the non-quantifiable realms of human experience. This 'life-world' with its messy ambiguities, irrational decision criteria and contingent histories is, for example, frequently relegated by model builders – even becoming parameterised in some models as a species of 'noise'. However, the omission of discussion on the moral and ethical basis of human-environmental problems can only succeed in further promoting a scientistic and technocratic discourse – one which believes that pollution, land degradation, coastal development etc. can be problematised within the conventional deductive methods of science, or rendered as a species of game theory with optimal solutions.

Before all else, problematising sustainability requires an acknowledgment that it is fundamentally about people; i.e. the capacity of social groups, not simply to survive, but to perpetuate themselves under conditions of food security and adequate welfare provision. This is pre-eminently a moral imperative that cuts across the neatly assembled packets of scientific knowledge and their instrumentalist projections within large-scale complex models.

4.1.3 Towards a Working Definition

While the search for a comprehensive definition of sustainability is destined to remain elusive, what is clear is that an important distinction must be made, as to which kind of sustainability we are dealing with – be it environmental, economic or social – since each has a distinctive meaning as well as being relative to a specific spatio-temporal domain.

But beyond the terminological confusion and slack usage, there are more fundamental problems which need to be addressed if we are ever to consider incorporating sustainability as an important and potentially useful tool. For example, with respect to watershed management issues, regardless of whether we are discussing resources, economies or societal systems, we must address a number of basic contextual questions:

1. Sustainable Watersheds for Whom?

It matters a great deal whether our target audience is regional water authorities, government agencies, industrialists, local communities, or indeed the individual farmer. The wide spectrum of interests, values and philosophies represented by these stakeholders demonstrates the futility of generalisation. Moreover, the ability to wield power (i.e. Who controls the water? Who owns the land?) has an important effect on which definition of sustainability will ultimately prevail.

2. Sustainable watersheds over what time period?
When speaking about a particular water resource, policy or type of socio-economic structure, the time span over which sustainability is envisaged is a critical factor – it matters a great deal whether we are talking about months, years, decades, or centuries.

3. Sustainable watersheds at what spatial scale?
In discussing ecological, social or economic processes, it needs to be established whether sustainability refers to geomorphological channel development, upstream, downstream or coastal delta regions. Within each of these levels are embedded local, regional and supra-regional vested interests identified by a variety of decision-making criteria – sustainability inevitably has a different meaning at each scalar level.

4. Sustainable watersheds for what purpose?
Political and economic organisations relative to different scales may achieve so-called sustainability, but at a cost to other humans; i.e. it may have significant ethical moral and welfare consequences for others. Whether water allocation policies preferentially favour tourism, commerce or agricultural needs, is ultimately related to policy decisions, which are themselves reflections of specific value systems and/or ideologies. In this sense, it is not the neutral category it is often portrayed as.

Thus, sustainability must not only be temporally and spatially defined, but most importantly, it must be contextualised with respect to specific political, ethical and social parameters. Unfortunately, as we have noted, the dominant model of western science within which the current sustainability discourse is situated, has meant that these contextual issues have been either poorly addressed, or relegated to secondary concerns.

4.1.4 Sustainability and Historical Knowledge

But there is another problem which needs to be addressed and that is the general *ahistorical* nature of the sustainability debate. If we are to learn anything of academic value about sustainability within the context of societal systems, then what is clear is that it must be studied from a *historical* perspective: people, institutions and the ecosystems they inhabit share one thing in common – they are all products of historical evolution. In a sense, the persistence of human societies is a consequence of their adaptive use of culture, which can be conceived as representing stocks of historically derived knowledge.

But this lack of historical perspective is best understood within the context of the dominant epistemology underpinning the current model of science. The contemporary debate within which sustainability is couched, is predicated on a model of knowledge that maintains a false separation between biophysical and societal phenomena – an expression of the erstwhile nature/culture dichotomy that has dominated western thought for centuries. While a number of integrated research programmes have attempted to tackle this problem – particularly in their critiques of reductionism and the poverty of neo-classical economics - little real progress has been made in recognising the need for a revised model of scientific enquiry, one which recognises the importance of establishing a dialogue between what have been perceived as mutually exclusive knowledge domains: *scientific* knowledge, *institutional* knowledge, *technical* knowledge and *local* knowledge (McGlade, 2001b). What we are arguing is that real insight into the nature of complex socio-economic systems and their dynamics cannot be understood, unless we are prepared to inscribe a new research territory, one in which a variety of knowledge domains can co-habit. Such a framework must additionally recognise the primacy of historical processes – both determined and contingent – for an understanding of the nonlinear dynamics, which articulate socio-economic systems.

By contrast to this model of knowledge acquisition, contemporary scientific research practice seems to have an aversion to history and its lessons. Indeed as Tainter (1995) points out, it is curious that in our modern problem-solving world we do not seek to utilise the vast data resources represented by the reservoirs of historical experience and knowledge. Typically, most policy makers are only interested in the recent past in their search for precedents. In addition, conventional research strategies tend to recognise the systemic nature at the expense of the historical component. Thus, while we have a greater opportunity than at any previous time in our history to understand the role of long-term processes in the creation of contemporary problems, this opportunity is largely ignored.

Whether it is simply a question of arrogance and a belief in the superiority of our 21st century scientific reasoning is not clear. Nevertheless, what we are arguing here is that historical knowledge is not only important, but a pre-requisite for understanding the nature, current status and future potential of complex socio-economic systems. Indeed, we might go further and claim that issues related to sustainability have no meaning if they are uncoupled from the larger long-term causalities of which they are an inevitable product. In short, history *matters*.

For example, with respect to the relationship between climate change and hydrology, Benito et al. (1996) have demonstrated the vital importance of understanding historical flood regimes as part of a long-term dynamic. Using historical data from the Iberian Peninsula, their analysis throws important light on the complex causalities linking hydrological responses to climate variability. Similarly, within the context of semi-arid watersheds, there is a great deal of latent information resident in time-series data sets relating to long-term drought and precipitation trends. What these data demonstrate is the highly unpredictable nature of climatic phenomena and the existence of discontinuous 'phase changes' in precipitation patterns – circumstances that further complicate our understanding of hydrological regimes and their effects on land-use activities. Under such conditions of uncertainty, there is a real need to understand the vulnerability and sensitivity of landscapes to change.

4.2 Landscape Sensitivity

The vulnerability of Euro-Mediterranean landscapes to change – and particularly their perceived sensitivity – has become a major research topic within the contemporary environment discourse. Paradoxically, while the term has wide usage (e.g. Thomas and Allison, 1993), a survey of extant environmental literature suggests that landscape sensitivity is an elastic term that can be moulded to suit a variety of contexts spanning land use change, geomorophological evolution, ecological succession dynamics, and/or the assessment of tourist carrying capacity (Goudie, 1986; Roberts, 1989; Naveh and Lieberman, 1990; Thomas and Allison, 1993).

In fact, landscape sensitivity as it is conventionally found in various aspects of the environmental and geographic literature, has been conventionally associated with geo-biophysical phenomena. Thus, among the most common applications are landscape assessments of geomorphic sensitivity – particularly in view of the wide spatial variation in the ability of landforms to incorporate change (Brunsden, 1990). By contrast, another definition of sensitivity emphasises the ability of landscapes to resist change (Brunsden and Thornes, 1977).

Conventional approaches seek to estimate the natural geomorphic sensitivity of landscapes and watersheds to a variety of land use activities (e.g. forestry, agriculture, urbanisation, tourism), which are characteristically viewed (and subsequently modelled) as 'disturbances'. In such studies, sensitivity analysis is designed to provide a quantifiable measure of the terrain's susceptibility to change (Turner and Gardner, 1991; Loh and Hsieh, 1995).

A key assumption here, is that landscapes can be classified in terms of their relative susceptibility to erosion, fire and landslide processes (perturbations), and therefore in their ability to cope with human imposed activities. This research orientation is, however, complicated by the fact that both hydrologic and geomorphic processes display wide variations in terms of their sensitivity to pollution generated by urban or industrial waste and/or the significant land use changes wrought by the growth of commercial and tourist construction projects. Nevertheless, predicting the probability of possible catastrophic change to river systems and flood regimes is a prominent research issue and this has led to the search for statistical indices of sensitivity, designed to help environmental managers with 'bottom line' scenarios from which they can anticipate future problems. Normally, GIS and Remote Sensing technologies are used in an effort to model the controls on soil erosion, vegetation growth, hillslope hydrology and water nutrient cycling. These studies are generally directed at evaluating potential land use change and include a variety of EC funded research programmes dealing with Mediterranean desertification (e.g. EFEDA, MEDALUS) and modelling landslides (NEWTECH). While these field-oriented, data-rich projects have been responsible for the collection of an important array of data sets on climate, soils, hydrological regimes and vegetation dynamics, as well as exploring functional responses of ecosystem variables, their *raison d'etre* has been that of physical measurement and prediction[3].

Beyond physical geography, perhaps one of the most common constructions of landscape sensitivity appears under the rubric of environmental impact assessment

[3] The goal of prediction is the *sine qua non* of our western society. However, despite a substantial body of research in a number of fields, demonstrating that our embeddedness in a complex evolutionary system *precludes* long-term prediction, much research still proceeds as though complexity resides in some hyper- theoretical arena; thus for all practical purposes it can be put to one side, while conventional analytical methods prevail.

(EIA) and related methodologies used to assemble environmental inventories and audits (e.g. Canter, 1996). Interaction matrices and checklists are common tools in these approaches and are used to generate quantitative estimates of the magnitude of potential impacts. Statistical summation of numerous attribute values forms the basis for insight into the future vulnerability and/or sustainability of the landscape (e.g. Canter, 1979, 1986; ESCAP, 1990).

These issues are particularly pertinent to fluvial systems and catchments generally. In fact, watersheds occupy a particular position with respect to issues of sensitivity, and this is most acute at the land/water interface in any catchment system. For example, the location of upstream activities such as agriculture and industrial processes means that river systems are constantly endangered by the threats posed by nitrates, as well as the variable water quality imposed by ground water pollution and changing urban/industrial and tourist demands. This has led some researchers to compute 'natural' watershed sensitivity as an estimation of a watershed's natural ability to absorb land use disturbance without unacceptably high level of impact (USDA, 1988).

However, the response of watersheds to disturbance events is in reality, complicated by the enormous variety of spatial and temporal ranges involved – spanning macro-scale climatic events all the way down to the micro-level dynamics of soil formation processes. In fact many watershed effects have characteristic substantial delays (often over years and decades) before their effects are manifest. Moreover, since the variables and turnover rates involved in river catchment processes differ from one watershed to another, this mitigates the construction of any generic model of watershed sensitivity (Newson, 1992).

Missing from many of these studies of landscape and watershed sensitivity, is a conception of the importance of scale. Importantly, landscape sensitivity issues are inevitably related to the temporal and spatial scale under investigation, rendering any scalar aggregation problematic. For example, behavioural aspects of geomorphological systems may be regarded as sensitive at one spatio-temporal scale but not at another. Understanding sensitivity is also made more difficult if we focus primarily on statistical approaches. These produce static descriptions of what are essentially *dynamic* processes; thus they misrepresent the inherent instability and nonlinear interactions that are the defining aspects of all complex socio-natural systems.

But perhaps the real weakness of the models discussed above concerns the way that they are frequently decoupled from human societal processes and especially the politics of management[4]. These latter are usually seen as the preserve of the social sciences and effectively relegated as problem sets for other disciplines. Perhaps the most problematic aspect of these studies is that anthropogenic factors are seen as external to the system; thus human intervention is modelled in terms of 'impacts' or 'perturbations' on the system[5].

---

[4] For an exception, see the ARCHAEOMEDES Programme (van der Leeuw 1998) which attempted to provide research contexts for the study of human-environment interaction.

[5] This is essentially an equilibrium view, with humans disturbing some hypothesised steady state to which the system aspires. In essence it perpetuates the age-old dichotomy, viewing the natural landscape as separate from the social and cultural realms.

4.2.1 Landscape Sensitivity Mapping

Attempts to go beyond the limitations of purely biophysical models have focused on technological advances provided by more sophisticated GIS and Remote Sensing technologies, and their ability to examine interactions between environmental, economic and social data sets (Arroyo-Bishop and Carlà, 1997; Schneider and Bartl, 1998). Additionally, the use of GIS systems has also enabled the assessment of potential impacts of land use strategies on the cultural landscape generally and specifically with reference to threatened heritage sites and monuments at a regional level (Palumbo and Powlesland, 1997; Hill and Aspinall, 1999). For example, McGlade et al. (1999)[6] developed an integrated Landscape Sensitivity Mapping System (LSM) to investigate the sensitivity of the cultural landscape to threats posed by the contested territorial claims of conservation, agriculture and tourism. This pilot study in the Emporda region of north-east Spain was concerned with isolating the primary drivers of change, with data sets partitioned into three analytical categories: 1) *ecological sensitivity measures*, 2) *economic sensitivity measures*, and 3) *socio-cultural sensitivity measures*. Spatially referenced data collected from these sectors were overlain to search for incompatibilities, discontinuities and correspondences based on a variety of different analytical criteria. The LSM system isolated and mapped the spatial distribution of key combinations of variables that act to create potentially vulnerable outcomes. Crucially, this methodology, was designed, not as an input/output system, focused on single answers – as with EIA and conventional landscape sensitivity methods – but rather, to promote a species of knowledge based system (KBS). The system was designed to act as a repository for different knowledge domains (environmental, economic, cultural) and to provide decision support material to help generate negotiated solutions to the contested issues that characterise multiple stakeholder landscapes.

4.2.2 Landscape Sensitivity: Some Representational Problems

Despite the continuing popularity of concepts such as 'sensitivity' and 'vulnerability' and, indeed, their centrality to environmental impact assessment programmes, they are basically inadequate as descriptors of complex socio-natural systems. A clear problem shared by most methodologies is the separation between the physical environment and what is perceived as a distinctive social and cultural environment.

However, it needs to be remembered that the physical environment has evolved in concert with (and as a product of) human action, forming a reciprocal socio-natural system (McGlade, 1995, 1999a, 2001a). Thus, any approach to landscape sensitivity that focuses exclusively on the biophysical aspects of the system (e.g. climate, geomorphological processes, hydrology etc.) is seriously incomplete as a represen-

[6] This study was carried out within the EC ARCHAEOMEDES Project, 1996–1999 (McGlade and Picazo 1999).

tation of the complexity of human-environment relations. The decoupling of the social and political underpinnings of the landscape to facilitate impact assessment checklists and quantitative model building, will only serve to generate fictive landscapes in which human action is ascribed the role of an *external* variable 'driving' the system, or 'impacting' on the environment. This commonly invoked model of humans as 'perturbations' casts them as somehow separate from the environment; resulting in spurious conceptualisations such as 'human impact'.

Ultimately, sensitivity is based on the potential and likely magnitude of change within the landscape system, as well as its ability to absorb perturbation. In short, the 'sensitivity' of a system to change induced by either biophysical or social phenomena is ultimately a function of its inherent *resilience*. Thus, what we shall argue is that attempts to understand sensitivity or vulnerability criteria through indices or other statistically derived criteria, will always be compromised. In essence, a more productive way forward is to situate such issues within a complex systems framework, so as to focus on one of the key aspects of sustainable systems, i.e. their *resilience*.

4.3 Resilience and Sensitivity

4.3.1 Resilience

Despite its frequent usage by ecologists, economists and some social scientists, resilience is not a unitary concept with precise and unambiguous definition. In the ecological literature, for example, it has two distinct meanings. The first emphasises stability, control and constancy (*engineering resilience*) – attributes of a desire for optimal performance, while the second, by contrast, focuses on persistence, adaptedness and unpredictability (*ecological resilience*) – attributes of a complex systems perspective. These latter are consistent with sustainability (Holling, 1996). Research using a model of engineering resilience, deals with stability near an equilibrium state and is concerned with resistance to disturbance and speed of return to equilibrium (e.g. De Angelis et al., 1980; Pimm, 1984; Tilman and Downing, 1994). By contrast, ecological resilience focuses on conditions far from equilibrium and is concerned with the role of instabilities in pushing the system beyond a threshold or bifurcation point, to a new stability domain. Here, resilience is measured by the magnitude of disturbance that can be absorbed before the system changes structure (Holling, 1973). A wide variety of applications exploring ecological resilience now exists, spanning resource ecology, wildlife management, fisheries, animal ecology and plant-vegetation dynamics (e.g. Holling, 1986; Walker et al., 1981; Walters, 1986; Sinclair et al., 1990; Dublin et al., 1990).

Studies such as these have been instrumental in shifting the ecological debate from a model based on the maintenance of stability, to one dominated by a sequence of interacting adaptive cycles based on a developmental sequence defined by four functions: exploitation, conservation, release and re-organisation (Holling, 1986). More recently, these ideas have been extended to encompass the idea of *panarchy*, which emphasises the evolutionary nature of nested adaptive cycles, with each level going through the cycle of growth, maturation, destruction and renewal (Gunderson et al., 1995). A key emphasis in this model is that periods of gradual growth and rapid transformation not only coexist, but act to complement one another (see also Günther and Folke, 1993).

4.3.2 *Resilience and Societal Systems*

All socio-economic systems seen to persist – particularly over long time periods – can be described as being characteristically resilient, in the sense that they are able to *incorporate* change and perturbation without collapsing. This ability to absorb changing circumstances as defined by environmental, social, political or cultural fluctuations is itself a function both of the flexibility of structural organisation and system history. The role of history is of crucial importance, in the sense that a particular regime that has been exposed to regular, periodic disturbance, will be more adapted to periodic change than a system which is visited by perturbation and/or extreme events on an irregular basis.

Any loss of resilience, will move a particular socio-economic system closer to unstable thresholds, causing it to flip from one attractor state to another (metastability); thus, for example, exploitation to extinction of a particular resource will have an effect on the local ecosystem, inducing system transformation and an irreversible change to an alternative state. Resilience can be said to bc one of the primary properties of nonlinear, nonequilibrium systems and needs to be understood more fully if we are to come to terms with sustainable social-natural systems.

A major problem that we are faced with in pursuit of a model of social-natural resilience within the context of watershed management, is that this cannot be deduced from conventional approaches to landscape sensitivity. However, neither can it be derived by the simple superimposition of Holling's (1986) resilience cycle for ecological dynamics. As we have already noted, this general theory of ecosystem function - incorporating insights from hierarchy theory (Allen and Starr, 1982; O'Neill et al., 1986) - has been argued as an appropriate basis for understanding the generic evolutionary behaviour underpinning ecological, economic and societal dynamics (e.g. Gunderson et al., 1995; Berkes and Folke, 1998; Peterson, 2000).

Notwithstanding the important insights that this evolutionary model provides, its essentially 'organic' nature is an inappropriate model for capturing the complexity of societal systems. In fact this organic formulation is consistent with a long philosophical tradition. For example classical authors [7] as well as early Christian writers [8]

[7] Polybius, writing in the 2nd century BC, when accounting for the defeat of Carthage by Rome noted: Every organism, every state and every activity passes through a natural cycle, first of growth, then of maturity and finally decay. Thus, at the time of their original conflict, Rome was in the ascending phase of the cycle, while Carthage was in decline.

[8] The third century Christian writer Cyprian in a passage quoted in Toynbee (1962) deplores the cycle of senescence and decay that can be seen in the world around him as part of the natural order of things.

emphasised the similarity between natural and societal dynamics, believing that societies could be understood by direct analogy with organisms, following a cycle of growth, maturity, senescence and death (Tainter, 2000). Such ideas, implicit in the later work of historians such as Oswald Spengler (1918) and Arnold Toynbee (1962) held sway in the social sciences until the 1980s when their structural shortcomings, particularly the underdeveloped relationship between agency and structure, as well as their inherent evolutionism, were critiqued by a number of sociologists and anthropologists, most notably Anthony Giddens. Importantly, Giddens (1979, 1984) provides a robust argument against the idea that societies 'adapt' to anything, since they are not equivalent to biological organisms (1979: 21). Instead, social change is seen as non-teleological – a set of contingent, discontinuous transitions which have no inherent developmental logic or pattern. However, despite these caveats, an increasing number of environmental scientists, resource managers and ecologists continue to apply ecosystem resilience ideas to socio-economic systems (e.g. Gunderson et al., 1995; Peterson, 2000).

An additional problem in utilising ecological resilience as an analogy for societal systems, is that human systems are not *neutral*; they are an historical product of specific social, political and cultural relations: a factor running all the way from local relations of production to larger scale regional, national and global levels of interaction. Thus, if we are to attempt to isolate the important driving forces of irreversible change which represent a non-sustainable option for society – then we must situate such goals within a milieu that recognises that (1) all landscapes (*sic* environment) are embedded in webs of power relations, and (2) these networks of power act to both enable and constrain human aspirations and desires. It is in the exercise of such power that the moral and ethical universe within which humans are situated, is subject to substantial modification and even destruction.

In summary, while resilience is a useful concept for understanding the long-term evolution of human-modified watersheds, it needs to be reconfigured to take account of the specific human and socio-political contexts that drive system transformation and change. In essence, we might summarise the main attributes of resilience from a socio-natural perspective as having the following characteristics:

5. The amount of re-organisation and change a social system can undergo, while still retaining the basic institutional and socio-economic structures
6. The degree to which the system's structure is capable of self-repair and self-organisation

This implies:

7. Institutional flexibility
8. The conscious use of historical knowledge
9. The desire to increase the capacity for knowledge production and learning
10. Conscious management of change to incorporate uncertainty and unintended
11. consequences

4.4 The River Catchment as a Complex System

In pursuit of a viable model of catchment management, we have already situated our discussion within the wider context of the sustainability discourse and identified a number of caveats with respect to landscape sensitivity and resilience criteria. In what follows we shall now attempt to incorporate the important attributes of a resilience perspective, for a better understanding of catchment systems viewed as nested sets of social, political, economic and environmental processes. Given the clear difficulties involved in such an enterprise, we shall begin logically by examining the primary attributes of a complexity perspective.

4.4.1 The Nature of Complexity

Recent years have seen the arrival of a new interdisciplinary approach to the analysis of complex, nonlinear systems and a gradual incorporation of these ideas into fields as far apart as chemistry, physics, ecology, urban and regional geography and the social sciences generally (Edmonds, 1996; Byrne, 1998). Complex systems are those systems 'whose aggregate behaviour is both due to, and gives rise to, multi-scale structural and dynamical patterns which are not inferable from a system description that spans only a narrow window of resolution' (Parrott and Kok, 2000).

As a new interdisciplinary field, Complexity Theory (Waldrop, 1992, Kauffman, 1993) is essentially concerned with studying the general attributes of nonlinear systems and exploring their propensity to follow unstable and chaotic trajectories (van der Leeuw and McGlade, 1997). Beginning in the early 1990s, this perspective and its central ideas has moved beyond the natural sciences to penetrate the social sciences, where complexity has been viewed as having potentially profound consequences for conventional epistemologies (Hayles, 1991; Byrne, 1998; Johnson, 2001).

Despite the diversity apparent in the complexity literature, there are however a number of key features that seem to be resident in all complex systems and are of central relevance to understanding the behaviour of river catchment systems. Among these are:temporal and spatial self-organisation, emergence, adaptivity, and critical levels of connectivity (Parrott and Kok, 2000). What we shall argue here is that a complexity perspective provides an appropriate context within which watershed dynamics – as a species of complex system – can profitably be analysed.

River catchments are complex and constantly evolving entities and like any 'moving target' they are difficult to analyse. This is rendered all the more problematic when we add the fact that their spatial development takes place at multi-scalar levels, as well as being articulated by a whole spectrum of different temporalities. Elsewhere (McGlade, 1999b), I have argued that the socio-natural world is defined by sets of distinctive temporalities that can be defined as *intrinsic times*; thus the biological, social, political and technological systems within which humans are situated can be characterised by inherent system times. These are the times that inhere in all social and natural activities – the turnover or reproduction times – extending from the reproduction of the cell, through plant and animal cycles to large scale glacial and planetary time scales (cf. Bender and Wellbery, 1991; Kummerer, 1996).

These intrinsic times and their spatial correlates, collectively form a nested spatial-temporal hierarchy. From our current perspective, recent research in the Vera Basin, south-east Spain, has demonstrated that multi-scalar temporalities act to structure these semi-arid environments (Fedoroff and Courty, 1995; McGlade, 1995). The primary message of this research is that landscape structure emerges as a result of the intersection of temporalities, ranging from the slowest processes such as tectonic movements (10^7), climatic cycles (10^5), all the way to population dynamics (10^2) and other micro-level phenomena (10^{-1}). Importantly, these temporalities are consistent with differential rates of change. Thus, we have slow, cumulative rates represented by glacial and tectonic movements, on which are superimposed annual and seasonal vegetational dynamics, along with micro-morphological soil structuring and intensive precipitation events ('gotas frias'). Research on a number of ecological systems shows that discontinuity – and frequently catastrophic outcomes – can be the result of the conjuncture of 'fast' and 'slow' variables (Holling, 1986). Such complexity is further enhanced by the superimposition of the array of time 'signatures' that characterise human social, political and economic systems. What we have in effect, are sets of *intertemporal dependencies*, defining a reciprocal dynamic that maps the social on to the natural and the natural on to the social. (McGlade, 1995, 1999b).

With respect to our focus on catchment systems, this emphasis on intrinsic times and their scalar attributes underlines the importance of studying complex processes, not simply in terms of change, but from a perspective that emphasises the role of self-reinforcing (positive feedback) processes in generating structure. Moreover, it is not change *per se* that is important, rather we must shift our focus to questions which deal with (i) the *rate* of change and perhaps, more important, (ii) the *changing* rate of change. It is these attributes which, above all, define the complex dynamics of socio-natural systems. It is within this specific context that we must place our research; that is with a view to understanding the relationships between climate variability, fluctuations in agricultural production, environmental pollution and management regimes. Watershed sensitivity is thus a complex systems concept.

In recent years, research on river systems has gradually moved from equilibrium ideas to a recognition that river dynamics are characteristically metastable, i.e. periods of apparent stability are interrupted by episodes of rapid change as the system moves to an alternative stability regime. This model of resilience is focused on the notion of thresholds (Newson, 1992:31) as underpinning river basin morphology, and replacing two dominant evolutionary approaches which have emphasised I) catastrophic change and II) gradualism, or slow progressive change. The inherent instability in the system is an endogenous source of change producing threshold phenomena. Threshold dynamics are observed in river basins:

(a) modified by artificial development, such as dams, irrigation projects or urbanisation;
(b) semi-arid river basins where sediment supply is affected by alternating drought and flood regimes. Basins are also affected by periodic fire.

It is in this sense that catchments and the nonlinear transformations in biophysical properties that they display, can be defined within the context of the model of resilience discussed above.

4.4.2 Catchment Systems and the Role of Climate

In addition to these endogenous sources of change, the sensitivity and vulnerability of watersheds – and their resilience – is also a function of exposure to specific climatic regimes. If we are to understand the resilience of watershed systems to climatic conditions, then this requires an initial classificatory distinction between;

(a) the sensitivity of hydrological systems to *climate change*, which is characteristically slow, and
(b) the sensitivity of hydrological systems to *climate variability*, which is, by comparison, relatively fast.

It is this second definition which is of primary concern for any research directed at the relationship between climate and watershed sensitivity. Knowledge of climate variability (Ruttenberg, 1981:27) suggests that three different types of information must be taken into account:

1. Normal, expected fluctuations around some mean value derived over a long period of climate history, which generally has a range which can be determined from long-term records. These types of fluctuations are generally oscillatory, but not cyclical
2. Rare and extreme events, such as frequent floods and prolonged droughts
3. Long-term events, such as cooling or warming periods which span a century or more. The importance of these in promoting significant social effects such as migration needs to be emphasised

In addition, when discussing the development of river systems with respect to climatic phenomena, we must account for the impact of shifts in global atmospheric circulation on river-flood behaviour. For example, it is well known that changes in the magnitude and frequency of extreme events are significant in terms of the impact of climate change on sediment and *P, N* and *C* transport in fluvial systems (Wasson, 1996). Most important, the interdependence which arises between seasonal flooding and agricultural production is vulnerable to shifts in the global atmospheric circulation, causing significant shifts in regional rainfall patterns; e.g. the *El Niño* event of 1983 was reflected in a substantially reduced flood peak on the lower Amazon, while records from the Parana river for the same period indicate that the same *El Niño* event was reflected in substantially increased flows (Vorosmarty et al., 1996). The more recent *El Niño* manifestation of 1997 has already demonstrated a variety of extreme weather conditions such as floods and catastrophic landslides, showing the rapid manner in which societies must adjust if they are to avoid potential disaster.

Generally speaking, we need to understand the role of long-term processes in generating social-natural dynamics. More specifically, the dynamics of long-term trends, upon which short-term fluctuations and responses are superimposed, can only be detected and quantified within a historical framework (Wasson, 1996). Most importantly, threshold phenomena change our perception of temporal scales over which change occurs, as well as providing an intrinsic source of change, i.e. it is not necessary to invoke external events such as climate or extreme floods.

Managerial and political dimensions

Hydrological and watershed sensitivity is, of course, not simply a function of climatic and biophysical processes. As we have noted earlier, the biophysical environment is but a small part of a much larger and more complicated equation, for the river basin, as a socio-natural product, is fundamentally and inescapably governed by political forces. These political forces, both intended and contingent, operate on a variety of local, regional and national scales. Thus, water utilisation patterns are characterised by cross-scale and inter-catchment administrative dynamics that are constantly in flux. This is as much through contested territorial issues promoted by commercial and urban stakeholders, as by changing tourist and conservation issues, and is particularly acute in semi-arid zones.

It is here that we encounter the various faces of legal, bureaucratic and managerial control over resources that constitute the sources of political power, particularly acute with respect to debates on ownership of water resources. One of the consequences of the reality of watershed systems as a series of contested space, is the need for appropriate policy exploration tools; indeed, this is a critical aspect of any watershed management initiative that aspires to integrative planning and sustainable outcomes. To this end, the need for decision-support tools is frequently argued, though the instrumentalist nature of many of these systems, while useful for exploring functional linkages between measurable variables, means that they are frequently found wanting when it comes to real-world policy exploration. The curious mix of determined and contingent processes that define societal systems is notoriously resistant to rule-based computational logic. Moreover, these difficulties are compounded by the realisation that large decision-support computer models will always find acceptance and implementation difficult within multi-user communities. They belong to a scientific discourse that by nature actively excludes large sectors of the community. For these reasons, the design of alternative conceptual frameworks of enquiry based on the interaction between knowledge communities, scenario construction and more democratic, inclusive participation, must be a major research priority.

4.5 Engaging the Past with the Present

4.5.1 The Lessons of History: Resilient Water Management Strategies

As we have already argued, much of the science directed at the management and resolution of water-use management conflicts seems to have little room for historical context. The persistence of a modernist scientific world view encourages the preference for engineering ('technofix') solutions to perceived problems; moreover, this is based on short-termism and is often a response to political pressure and the need to create demonstrable solutions. For these reasons, exposing the reality and intractability of many human-induced problems is not an option – preaching on the need to understand complexity is often dismissed as obfuscation and, more importantly, it does not win votes.

But history, as we have seen, has much to teach us, not least of which are some important lessons relating to the maintenance of resilient water management systems. Historically, the river catchment, seen as the locus of the control and management of water, flood plain and irrigation management has played a vital role in the resilience and long-term persistence of human communities. The ancient civilisations of Mesopotamia, Egypt, China and Peru, for example, were founded upon the management of water resources, and archaeology, as well as historical texts, has furnished us with remarkable evidence of the ability of water control systems such as irrigation, to sustain large urban civilisations. This evidence became the basis of Wittfogel's (1957) famous hydraulic hypothesis for the origin of the state, according to which the control of water requires an elaborate system of management and mass labour – something that can only be achieved by centralised power.

However, a variety of anthropological studies have demonstrated that the structures involved in irrigation agriculture cover a wide spectrum of social and political organisational types (e.g. Carneiro, 1970). Indeed, a number of successful systems are characterised by decentralised social structures, for example, the medieval huertas in Valencia (Glick, 1970) provide an outstanding case of a resilient water management system. In fact these huerta systems, which can be traced back to the Islamic period, were self-governing communes that were only disrupted during the period of 19th century industrialisation and agrarian reform; thus, they persisted for more than a thousand years – the social and political organisation of the Commons remaining relatively unchanged. In this sense, it is interesting to note that the arrival of feudal modes of production had little effect on the huerta systems, so thoroughly were they embedded, both culturally and economically, in society. With respect to our current discussion, here we have, by definition, a resilient socio-economic system that clearly merits more study.

So total is our focus on the 'now', the immediate present, that the historical context of events is lost, or at least disconnected, from the orbit of decision-makers and their knowledge systems. In fact, it is this very disarticulation that is at least partly responsible for the loss of resilience and collapse of many human-modified systems across the globe. A critical omission in this latter respect has been the continuing devaluation of systems of indigenous knowledge as being non-scientific, and hence of only anecdotal value.

For example, the Inca were skilled farmers – their irrigation systems have been mapped and archaeological research has revealed that they were involved, if not consciously, in the creation of sustainable environments then, at least, focused on the creation of resilient agricultural systems. Their sophisticated terraced irrigation systems produced biannual cropping of maize and potatoes (as reported by a native Andean, Felipe Guaman Poma in 1613). Remarkably, their stone wall terracing structures and canal systems endured due to an elaborate system of delegating maintenance activities within the community. Significantly, what we would refer to as 'economic' or 'subsistence' systems, were indivisible from the social and religious mores of the society. Once again, as with the Valencian systems, it should be of no little interest to us that for these reasons, irrigation systems lasted for up to one thousand years (Kendall, 1997:4).

The need for a long term perspective is particularly important if we are to gain an understanding of the wider historical compass; i.e. the dynamic which defines human intervention in soil, vegetation and hydrological cycles. From our current watershed perspective, we can thus usefully ask, what are the lessons of the past? Can we identify specific ways in which water control and management have contributed to societal persistence and/or collapse? Clearly a variety of evidence can be brought to bear on this question from the early hydraulic civilisations as well as examples from the Roman Empire. Newson (1992) has provided a summary drawn from a number of well-known archaeological examples:

- Social aspects of co-ordination and control have been as important – if not more so – than technological aspects (the Sumerian lesson)
- Distributional aspects of water management (e.g. irrigation, drainage, and flood control) can produce highly efficient developments (the Roman example)
- The fundamental legal principles upon which a society bases its approach to water management will powerfully influence and constrain the environmental outcome
- Scale issues are critical (the Indus lesson) because they control the distribution of information in the system, both technological and social

4.6 Conclusions

In line with our previous discussion of resilience that stressed the adoption of cultural mechanisms for gathering, storing and evaluating information, one way to view the river basin from a complexity perspective is to see it as a *knowledge mosaic*, enabled and constrained by a variety of political and economic organisational forms. Over time the connectivities between the institutional, administrative and scientific knowledge domains allow the flow of information and energy to generate emergent self-organising structure. Thus, what we can say, is that a measure of the resilience of a catchment is the degree of congruence between knowledge domains: scientific, institutional, technological and local. This implies convergence at a number of scales as well as administrative agreements across political and planning boundaries. Without such cross-scale institutional cooperation, conflict can inevitably arise, as a consequence of fragmented planning and legislation.

From a sustainability perspective, problems arise when there is a mismatch between knowledge domains, in effect a form of *cognitive dissonance* – effectively the rate of divergence between knowledge categories. Given the importance of the various legal, administrative and institutional processes involved in watershed development, we can usefully talk of *institutional resilience* and its analysis as a viable research goal.

Acknowledgments This work was financed by the European Commission under the AQUADAPT project (EVK1-CT-2001-00104).

References

Allen, T.F.H. and Starr, T.B. (1982). *Hierarchy: Perspectives for Ecological Complexity*. University of Chicago Press, Chicago.

Arroyo-Bishop, D. and Carlà, R. (eds). (1997). Remote Sensing for Geography, Geology, Land Planning and Cultural Heritage. *Proceedings of European Series*, 2960.

Bender, J. and D.E. Wellbery (eds.). (1991). *Chronotypes: The Construction of Time*, Stanford, Stanford University Press.

Benito, G., Machado, M.J., and Perez-Gonzalez, A. (1996). Climate change and flood sensitivity in Spain. In J. Branson, J. Brown and K.J. Gregory (eds) Global Continental Changes: The Context of Palaeohydrology. Geological Society Special Publications, p.115, pp.85–98.

Berkes, F. and C. Folke. (1998). *Linking Ecological and Social Systems*, Cambridge University Press, Cambridge, UK.

Breton, F. (1996). El littoral: bases per al planajemente i la gestió integrada d'un espai dinàmic i vulnerable. *Quaderns d'Ecologia Aplicada*, 13: 45–100.

Breton, F., Clapes, J., Marques, A., and Priestley, G. (1996). The recreational use of beaches and consequences for the development of new trends in management: the case of the beaches of metropolitan Barcelona (Catalonia, Spain). *Ocean and Coastal Management*, 32(3): 153–180.

Breton, F. and Sauri, D. (1997). Towards a redefinition of resources and hazards in Coastal Management examples from the lowlands areas of Catalonia Spain. *Coastal Management*, 25: 263–385.

Brunsden, D. and Thornes, J.B. (1977). Landscape sensitivity and change. *Transactions of the Institute of British Geographers*, 4: 463–484.

Brunsden, D. (1990). Tablets of stone: towards the ten commandments of geomorphology. *Zeitschrift fur Geomorphologie, Suppl.* 79: 1–37.

Bulloch, J. and Darwish, A. (1993). *Water Wars: Coming Conflict in the Middle East*, Victor Gollancz, London.

Byrne, D. (1998). *Complexity Theory and the Social Sciences*, Routledge, London.

Canter, L.W. (1979). *Water Resources Assessment – Methodology and Technology Sourcebook*. Ann Arbor Science Publishers, Ann Arbor, MI.

Canter, L.W. (1996). *Environmental Impact Assessment*, McGraw-Hill, New York.

Carneiro, R.L. (1970). A theory of the origin of the state. *Science* 169: 733–738.

Daly, H. (1994). Operationalizing Sustainable Development by Investing in Natural Capital. In A.M. Jansson, M. Hammer, C. Folke and R. Costanza (eds) *Investing in Natural Capital: The Ecological Economics Approach to Sustainability*, Island Press, Washington, DC, pp. 22–37.

De Angelis, D.L., Post, W.M. and Travis, C.C. (1980). *Positive Feedback in Natural Systems*, Springer, New York.

Dublin, H.T., Sinclair, A.R.E. and McGlade, J. (1990). Elephants and fire as causes of multiple stable states in the Serengheti-Mara woodlands. *Journal of Animal Ecology* 59: 1147–1156.

Edmonds, B. (1996). What is Complexity? In F. Heylighen and D. Aerts (eds.) *The Evolution of Complexity*, Kluwer, Dordrecht, The Netherlands.

ESCAP (Economic and Social Commission for Asia and the Pacific) (1990). *Environmental Impact Assessment – Guidelines for Water Resources Development*, ST/ESCAP/786, United Nations, New York, pp. 19–48.

Faucheux, S. & O'Connor, M. (1998). *Valuation for Sustainable Development - Methods and Policy Indicators. Advances in Ecological Economics*, Edward Elgar, Cheltenham, UK.

Falkenmark, M. and Lindh, G. (1993). Water and economic development. In P.H. Faucheux, S. and O'Connor, M. (eds). (1998). *Valuation for Sustainable Development: Methods and Policy Indicators*, Edward Elgar, Cheltenham, UK.

Fedoroff, N. and Courty, M.A. (1995). Le Role Respectif des Facteurs Anthropiques et Naturels dans la Dynamique Actuelle et Passée des Paysages Méditerranéens: Cas du Bassin de Vera, Sud-est de l'Espagne. In S.E. van der Leeuw (ed.), *L'homme et la Degrédation de L' environnement*, Juan Les Pins: Editions APDCA, pp. 339–356.

Giddens, A. (1979). *Central Problems in Social Theory*, Macmillan, London.

Giddens, A. (1984). *The Constitution of Society*, Polity Press, Cambridge.

Glick, T.F. (1970). *Irrigation and Society in Medieval Valencia*, The Belknap Press, Cambridge, MA.

Goudie, A.S. (1986). *The Human Impact on the Natural Environment*, Blackwell, Oxford.

Gunderson, L., Holling, C.S. and Light, S. (eds). (1995). *Barriers and Bridges to the Renewal of Ecosystems and Institutions*, Columbia University Press, New York.

Gunther, F. and Folke, C. (1993). Characteristics of nested living systems. *Journal of Biological Systems* 1: 257–274.

Hayles, N.K. (1991). *Chaos and Order*, University of Chicago Press, Chicago.

Hill, M.J. and Aspinall, R.J. (eds). (1999). *Spatial Information for Land Use Management*, Gordon & Breach, London.

Holling, C.S. (1973). Resilience and stability in ecological systems. *Annual Review of Ecology and Systematics*. 4: 2–23.

Holling, C.S. (1986). The resilience of ecosystems: local surprise and global change. In W.C. Clarke and R.E. Munn (eds) *Sustainable Development of the Biosphere*, Cambridge University Press, Cambridge, pp. 292–317.

Holling, C.S. (1996). Engineering resilience *vs*. ecological resilience. In P.C. Schutze (ed) *Engineering within Ecological Constraints*, National Academy Press, Washington DC., pp. 32–43.

IUCN/UNEP/WWF. (1980). *World Conservation Strategy: Living Resource Conservation for Sustainable Development*, IUCN, Gland, Switzerland.

Johnson, S. (2001). *Emergence: The Connected Lives of Ants, Brains, Cities and Software*, Allen Lane, The Penguin Press, London.

Kauffman, S.A. (1993). *The Origins of Order: Self-organization and Selection in Evolution*, Oxford Univesity Press, Oxford.

Kendall, A. (ed) (1997). *The Restoration of Pre-Hispanic Agricultural Systems*, Amauta Press, Cuzco, Peru.

Kümmerer, K. (1996). The Ecological Impact of Time. *Time and Society* 5: 209–235.

Loh, D.K., Hsieh, Y.T.C. (1995). Incorporating Rule-Based Reasoning In the Spatial Modeling of Succession In A Savanna Landscape. *AI Applications*, 1995, 9(1): 29–40.

McGlade, J. (1995). Archaeology and the ecodynamics of human-modified landscapes. *Antiquity* 69: 113–132.

McGlade, J. (2001a). Archaeology, Integrated Management and Contested landscapes: Towards Sustainable Solutions. In A. Amato (ed.). *Archaeological Sites, Territory and Local Development*, Naples, IMED Conference Proceedings, Rome, Italy.

McGlade, J. (2001b). *Complexity, Representation and Model-building: Towards a Dialogic Approach.* NEXSUS working paper, Brookfield Manor, Derbyshire, April 2001.

McGlade, J., Nazzar, A., Serra, P., Romagosa, F., Breton, F. and Verhagen, P. (1999). Assessing the vulnerability of the landscapes of the Emporda. In J. McGlade and M. Picazo (eds). *Human Ecodynamics and Land Use Conflict: Monitoring the Degradation-Sensitive Environments of the Emporda, north-east Spain.* ARCHAEOMEDES II final report. DGXII Environment and Climate Programme, Commission of the European Communities, Brussels, pp. 217–276.

McGlade, J. (1999a). Archaeology and the evolution of cultural landscapes: towards an interdisciplinary research agenda. In R. Layton and P.J. Ucko (eds). *Anthropological and Archaeological Approaches to Landscape*, Routledge, London.

McGlade, J. (1999b). The times of history: archaeology, narrative and nonlinear causality. In T. Murray (ed) *Time and Archaeology*, Routledge, London, pp. 139–163.

Naveh, Z. and Lieberman, A.S. (1990). *Landscape Ecology*, Springer, Berlin.

Newson, M. (1992). *Land, Water and Development*, Routledge, London.

O'Neill, R.V., De Angelis, D.L., Waide, J.B. and Allen, T.F.H. (1986). *A Hierarchical Concept of Ecosystems*, Princeton University Press, Princeton, NJ.

Pearson, M. and Sullivan, S. (1995). *Looking After Heritage Places*, Melbourne University Press, Melbourne.

Palumbo, G. and Powlesland, D. (1997). Remote Sensing and Geographic Database Management Systems: applications for the Protection and Conservation of Cultural Heritage. In D. Arroyo-Bishop, R. Carlà *et al.* (eds.). *Remote Sensing for Geography, Geology, Land Planning and Cultural Heritage. Proceedings. Euro. Ser.* 2960: 120–129.

Parrott, L. and Kok, R. (2000). Incorporating complexity in ecosystem modelling. *Complexity International*, 7: 1–19.

Peterson, G. (2000). Political Ecology and ecological resilience: an integration of human and ecological dynamics. *Ecological Economics*, 35: 323–336.

Pimm, S.L. (1984). The complexity and stability of ecosystems. *Nature*, 307: 322–326.

Roberts, N. (1989). *The Holocene: an Environmental History*, Blackwell, Oxford.

Ruttenberg, S. (1981). Climate, food and society. In Slater, L. E., and Levin, S. K. (Eds.). *Climate's Impact on Food Supplies*, pp. 23–37. Westview Press, Boulder, CO.

Schneider, W. and Bartl, R. (1998). Land Cover Mapping from Optical Satellite Images Employing Subpixel Segmentation and Radiometric Calibration. In Kanellopoulos, Wilkinson, Moons (eds.), *Machine Vision and Advanced Image Processing in Remote Sensing*, Proc. MAVIRIC Workshop, Kingston University, UK, Springer pp. 229–237.

Sinclair, A.R.E., Olsen, P.D. and Redhead, T.D. (1990). Can predators regulate small mammal populations? Evidence from house mouse outbreaks in Australia. *Oikos* 59: 382–392.

Smith, B. (1997). Water: a critical resource. In R. King, L. Proudfoot and B. Smith (eds) *The Mediterranean: Environment and Society*, Arnold, London, pp. 227–251.

Spengler, O. (1918). *The Decline of the West: perspectives on world history*, Harvard University Press, New Haven, CT.

Tainter, J.A. (1995). Sustainability of Complex Societies. *Futures* 27: 397–407.

Tainter, J.A. (2000). Global Change, History and Sustainability. In R.J. McIntosh, J.A. Tainter and S. McIntosh (eds). *The Way the Wind Blows Climate, History and Human Action*, Columbia University Press, New York, pp. 331–356.

Thomas, D.S.G. and Allison, R.J. (1993). *Landscape Sensitivity*, Wiley, Chichester, UK.

Tilman, D. and Downing, J.A. (1994). Biodiversity and stability in grasslands. *Nature*, 367: 363–365.

Toynbee, A.J. (1962). *A Study of History*, Oxford University Press, Oxford.

Turner, M.G. and Gardner, R.H. (1991). *Quantitative Methods in Landscape Ecology*, Springer, New York.

USDA Forest Service. (1988). Cumulative off-site watershed effects analysis. In: *Soil and Water Conservation Handbook* (ch. 20), Region 5, Pacific Southwest Region, San Francisco, California.

van der Leeuw SE. (1998). The ARCHAEOMEDES Project B: Understanding the natural and anthropogenic causes of land degradation and desertification in the Mediterranean basin. : Office of Publications of the European Union,Luxembourg,

van der Leeuw, S.E. and McGlade, J. (eds). (1997). *Time, Process and Structured Transformation in Archaeology*. Routledge, London.

Vörösmarty, C. J., Willmott, C.J., Choudhury, B.J., Schloss, A.L., Stearns, T.K., Robeson, S.M., and T. J. Dorman, (1996): Analyzing the discharge regime of a large tropical river through remote sensing, ground-based climatic data, and modeling. *Water Resouces Research*, 32: 3137–3150.

Waldrop, M. (1992). *Complexity: The Emerging Science at the Edge of Order and Chao*s, Simon & Schuster, New York.

Walker, B.H., Ludwig, D., Holling, C.S. and Peterman, R.M. (1981). Stability of semi-arid grazing systems. *Journal of Ecology* 69: 473–498.

Walters, C.J. (1986). *Adaptive Management of Renewable Resources*, Macmillan, New York.

Wasson, R.J. (1996), (ed). *Land Use and Climate Impacts on Fluvial Systems during the Period of Agriculture*, PAGES workshop report series 96(2).

WCED (World Commission on Environment and Development). (1987). *Our Common Future*, Oxford University Press, Oxford.

Wittfogel, K. (1957). *Oriental Despotism*, Harvard University Press, New Haven, CT.

Chapter 5
Using Economic Valuation Techniques to Inform Water Resources Management in the Southern European, Mediterranean and Developing Countries: A Survey and Critical Appraisal of Available Techniques

Ekin Birol, Phoebe Koundouri, and Yiannis Kountouris

5.1 Introduction

Water resources include surface water, groundwater, inland water, rivers, lakes, transitional waters, coastal waters and aquifers (Chave, 2001). Together, these water resources are crucial to human health and the natural environment, and are vital to any economy in the world. Water resources are necessary inputs to production in economic sectors such as agriculture (arable and non-arable land, aquaculture, commercial fishing, and forestry), industry (e.g. power generation) and tourism, as well as to household consumption (UNEP, 2005).

Over time however, water resources have been degraded and depleted globally. With respect to water quantity, these trends have grown stronger within the past century during which global freshwater-use increased sixfold, and 50% of global wetlands were lost (IUCN, 2005). In Southern European and Mediterranean countries statistics reveal significant water stress problems regarding water quantity and quality and report deterioration of the state of water resources during recent years. An example of the increased pressure on water resources, is given by the European Commission that reports a 20% increase in the area of irrigated land in Southern Europe since 1985 (EC, 2002). In Chapter 2 of this volume, Barraqué et al. report evidence from fifteen sites in six southern European countries, which face severe water stress in terms of water availability due to climatic conditions and increased demand patterns. Regarding water quality, their findings reveal the deteriorating conditions of groundwater stock, as 7 out of 15 sites exhibit below average groundwater quality. The availability and quality of water resources in the Mediterranean islands is further deteriorated due to their isolated nature and their inability to draw water from more distant resources. In Chapter 3 of this volume, Donta et al summarise the findings of the MEDIS project displaying the strain placed on water resources from increasing demand and natural causes.

These adverse effects on water are a result of increasing water demand from agriculture, industry, tourism, hydroelectric generation, as well as continued pollution.

P. Koundouri (ed.), *Coping with Water Deficiency*, 135–155.

The effects are further exacerbated by population growth, rapid urbanisation and climate change (UNEP, 2000). From an economic perspective, water resources are over-extracted and are not efficiently allocated. This is due in part to the existence of market and government failures at the local, national and international level. Private costs and benefits diverge from social costs and benefits, leading to social welfare losses (Pearce and Turner, 1990).

In recognition of the deterioration in the quantity and quality of water, several initiatives have been undertaken to ensure the sustainable management and conservation of this valuable resource. The EU's WFD aims to protect and achieve a "good status" for all water resources by 2015, with a combined approach of emission limit values, quality standards, and the introduction of more efficient water prices. There are also international efforts to conserve water resources, such as the 1971 Ramsar Convention on Wetlands of International Importance, providing a framework for national action and international cooperation for the conservation and wise use of wetlands (Ramsar, 1996).

The aims of this chapter are to highlight the need for economic analysis in the design and implementation of efficient and effective water resources management strategies and policies; to explain and critically assess the suitability of various economic valuation techniques for this purpose; and finally through a comprehensive review of the literature, to demonstrate how these methods can be used in the development of appropriate policies for sustainable water resources management in the Mediterranean and Southern European countries. The chapter is structured as follows: The next section discusses the role of economic analysis in efficient water resources management. In Sections 5.3 and 5.4, the most commonly used economic valuation methods, namely revealed preference methods and stated preference methods, are described. The context in which each of these methods can be used and their respective limitations are explained. The theory is illustrated with examples of existing studies from Mediterranean, Southern European and Developing countries that have employed these methods to estimate the values of water resources. Finally Section 5.5 concludes and discusses implications for water resources policy in Southern European, Mediterranean and Developing countries.

5.2 The Economics of Water Resource Depletion and Degradation: A Conceptual Framework

Even though water resources are vital for the functioning of any economy, they continue to be depleted and degraded at an unsustainable rate. This is true for both developed and developing countries alike, and is due to the nature of the economic development and growth path that has been chosen thus far, which has readily substituted environmental resources (such as water) for other forms of economic resources such as capital and labour for the production of goods and services that are deemed to be more productive and yield higher returns (Swanson and Johnston, 1999). This path has been chosen because the value of environmental resources has often been over-looked in development decisions. Economic efficiency occurs at the point where net social benefits (i.e., benefits minus costs) of an economic activity are maximised, or equivalently, when the marginal benefits are equal to marginal costs. To implement the most efficient social and economic policies that prevent the excessive degradation and depletion of environmental resources, it is necessary to establish their full value, and to incorporate this into private and public decision-making processes.

A widely accepted and often used framework for decision making is Cost Benefit Analysis (CBA). CBA is an analytical tool based in welfare theory, which is conducted by aggregating the total costs and benefits of a project or policy over both space and time (Hanley and Spash, 1995). A project or policy represents a welfare improvement only if the benefits net of costs are positive. Different management options will yield varying net benefits and the option with the highest net benefits is the preferred or optimal one.

A CBA of a policy or project with environmental impacts is complicated because many environmental resources (including most water resources) are public goods. A good is public to the extent that consumption of it is non-rival and non-excludable; It is non-rival if one person's consumption of the good does not reduce the amount available to others and non-excludable if it is possible to supply the good to everyone. Pure public goods cannot be provided by the price mechanism because producers cannot withhold the good for non-payment, and since there is no way of measuring how much a person consumes, there is no basis for establishing a market price. Public goods are therefore not traded in markets as private goods are, and are thus often under-produced or over-exploited by the market. This phenomenon is termed 'market failure' in economic terms. Both surface water and groundwater have public good characteristics in that people who extract them and use them are not paying their scarcity rents (both in terms of quality and quantity); they only pay the private extraction costs. When scarcity rents go unrecognised, the result is inefficiently high extraction or pollution rate over time and space (Koundouri, 2000). Other causes of market failure include insufficient or non-existent property rights, externalities, the lack of perfect competition (e.g., market power) and lack of perfect information. The property rights issue is especially important in the context of water resource management. If there were private property rights, then for example an upstream polluter of water would be legally required to compensate the downstream property rights owner for damages, thus leading to the 'optimal' level of pollution. Externalities refer to costs or benefits borne by individuals who are not directly involved in a market transaction, and who have not been compensated. Where market failures exist, government must intervene to allocate the resources efficiently. Generally, governments do not intervene to correct these failures because environmental conservation is not a high priority. In the case of water supply, a basic human necessity, the government has a stronger incentive to intervene to provide the population with clean water. Though this is true for both developed and developing countries, water quality standards in developing countries tend to be lower than in the developed countries (e.g., EU standards for drinking water quality are stricter than those of the World Health Organisation), and government intervention in the developing world is often slower due to budget constraints and incomplete or non-existent infrastructure and institutions. In addition, certain government policies such as subsidies, distort the prices of environmental resources thereby not accounting for their economic scarcity. These result in the phenomenon of 'government failure'.

To correct for these failures, the value of all the benefits provided by environmental resources need to be captured. Environmental economists have been at the forefront arguing that individuals may derive values from non-market goods, especially environmental resources, through many more sources than just direct consumption (Pearce and Turner, 1990). More specifically, they refer to the importance of considering the Total Economic Value (TEV) of an environmental resource. TEV recognises two basic distinctions between the value that individuals derive from using the environmental resources, i.e., use values, and the value that individuals derive from the environmental resource even if they themselves do not use it, i.e., non-use values. Use values can be further classified into three broad categories: Direct use values, indirect use values, and option values. Direct use values come from the consumptive use of the environmental resource itself. With regard to water resources, these include drinking water, irrigation, or as an industrial input (Table 5.1). For most private (normal) goods, value is almost entirely derived from their direct use. Many environmental resources however, perform an array of functions that benefit individuals indirectly: indirect use values of water resources include benefits such as flood control, nutrient retention, and storm protection. Finally, option value recognises that individuals who do not presently use a resource may still value the option of using it in the future. The option value for water resources therefore represents their potential to provide economic benefits to human society in the future.

A further major expansion of value of an environmental resource is the inclusion of non-use values (Krutilla, 1967). These are values that individuals may derive from environmental resources without ever personally using or intending to use them. These can be further classified into three categories, namely existence value, bequest value, and altruistic value.

Existence value refers to the value individuals may place upon the conservation of an environmental resource, which will never be directly used by themselves or by future generations. Individuals may value the fact that future generations will have the opportunity to enjoy an environmental resource, in which case they might express a bequest value. Finally, altruistic value states that even if the individuals themselves may not use or intend to use the environmental resource themselves, they may still be concerned that the environmental good in question should still be available to others in the current generation.

These concepts are illustrated in Fig. 5 1. The MNPB curve represents the marginal net private benefits of using water resources, where $MNPB_S$ curve represents the marginal net private benefits of using water resources exacerbated by subsidies to their use. The MEC_L is the marginal external costs borne locally from use of water resources and the MEC_{L+G} is the local and global marginal external costs from use of water resources, measured by the TEV of the water resources. These curves

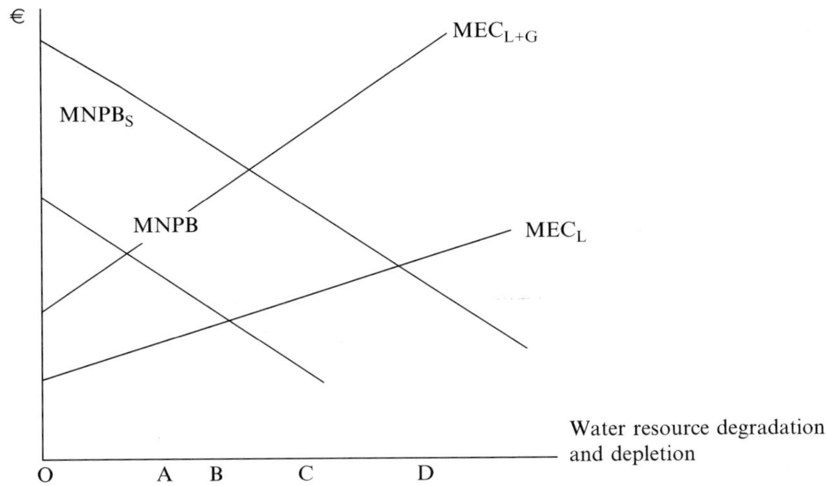

Fig. 5.1 Impacts of market and government failure and population growth on water use Source: Adopted from Pearce (2001); x axis is the decline in quantity and quality of water; y axis is the monetary costs and benefits

result in four equilibria, with four levels of water resource use. Point C is the local private optimum, where all externalities are disregarded and there are no subsidies to water use. Externalities are defined as benefits or costs, generated as by-products of an economic activity that do not accrue to the parties involved in the activity. An externality can be local, in which case it is confined to a specific location, or global, and it can be positive or negative.

Point D is the local private optimum, where, again all externalities are disregarded and water use is subsidised. Point B is the local social optimum, where local externalities are internalised but global externalities are ignored, and point A is the global social optimum, where all externalities are internalised. When an externality is internalised, the market and government failures have been corrected to the point where economic efficiency has been attained. The government failure is measured by distance CD, i.e., the quantity and quality of water resources that is lost due to its conversion for use in economic activities (e.g., irrigation for agriculture or a waste sink for pollution run-off from industry) as a result of government subsidies. Local market failure is measured by BC, and global market failure by AB. The distance AD reflects the inefficiency of water resource use, as shown by the divergence between the private and social optimum. The efficient use of water resources occurs at OA (Pearce, 2001).

To summarise, values of water resources are not straightforward to estimate for CBA purposes. This is not only because many of the water resources are public goods in nature, and hence do not have readily available monetary values attached to them, but also because their value is more complex compared to private goods. This complexity arises from the fact that the value of water resources are composed of both use and non-use values. Capturing the TEV of water resources is crucial to policy and management decisions because they can guide resource allocations among water resource conservation and sustainable management and other socially valuable endeavours, as well as within water resources, thus enabling society to allocate its scarce economic and environmental resources efficiently. Establishing the TEV would also assist in the design of economic incentives and institutional arrangements, and help to identify potential gainers and losers from current depletion and degradation of water resources (Drucker et al., 2001).

Various economic methods have been developed to capture the TEV of environmental resources. Table 5.1 lists the main economic methods that can be used to estimate the values of water resources. The advantages and disadvantages of each method, along with their uses in capturing the value of water resources, is the subject of the subsequent two sections.

Table 5.1 Components of TEV of water resources and appropriate economic valuation methods

TEV Component	Economic valuation methods*
Direct use values	
Irrigation for agriculture	PF, NFI, RC, MP
Domestic and industrial water supply	PF, NFI, RC, MP
Energy resources (hydro-electric, fuel, wood, peat)	MP
Transport and navigation	MP
Recreation/amenity	HP, TC, CVM, CEM
Wildlife harvesting	MP
Indirect use values	
Nutrient retention	RC, COI
Pollution abatement	RC, COI
Flood control and protection	RC, MP
Storm protection	RC, PF
External eco-system support	RC, PF
Micro-climatic stabilisation	PF
Reduced global warming	RC
Shoreline stabilisation	RC
Soil erosion control	PF, RC
Option values	
Potential future uses of direct and indirect uses	CVM, CEM
Future value of information of biodiversity	CVM, CEM
Non-use values	
Biodiversity	CVM, CEM
Cultural heritage	CVM, CE
Bequest, existence and altruistic values	CVM, CE

Source: With modifications adopted from Barbier (1991, 1997), Woodward and Wui (2001), Brouwer et al., (2003), and Brander et al., (2006).

*Acronyms refer to Production Function (PF), Net Factor Income (NFI), Replacement Cost (RC), Market Prices (MP), Cost of Illness (COI), Travel Cost Method (TCM), Hedonic Pricing Method (HP), Contingent Valuation Method (CVM), and Choice Experiment Method (CEM).

5.3 Revealed Preference Methods

Revealed preference methods, also known as indirect valuation methods, look for related or surrogate markets in which the environmental good is implicitly traded, i.e., if it is one of the many components of a good that is purchased by the consumer (Lancaster, 1966). Information derived from observed behaviour in the surrogate markets is used to estimate willingness to pay (WTP), which represents individual's valuation of, or the benefits derived from, the environmental resource. Two such methods prevalent in the environmental economics literature are the hedonic pricing and the travel cost methods. These methods are suitable for valuing those water resources that are marketed indirectly and are thus only able to estimate their use (direct and indirect) values.

5.3.1 Hedonic Pricing Method

The hedonic pricing method (HPM) is based on Lancaster's characteristics theory of value (Lancaster, 1966), which states that any good can be described as a bundle of characteristics and the levels these take, and that the price of the good depends on these characteristics and their respective levels. It is commonly applied to variations in housing prices that reflect the value of local environmental resources. The price of a house will reflect its relevant characteristics, i.e., number of bedrooms, number of bathrooms, size, schools in the neighbourhood, level of crime, etc., in addition to the local environmental resources such as ambient air quality, noise levels, aesthetic views, water quantity or quantity.

It follows that an implicit price exists for each of the characteristics and an implicit marginal WTP, which represents an individual's valuation of the incremental unit of the environmental resource can be identified statistically. A limitation of the HPM is that it only measures direct use values of water resources as perceived by the consumers' of the good in which it is implicitly traded. Services such as flood control, water quality improvement, habitat provision for species, and groundwater recharge may provide values that benefit individuals far away, beyond the consumers of the good, which the HPM is unable to capture (Boyer and Polasky, 2004).

The HPM was developed by Griliches (1971) to estimate the value of quality change in consumer goods. The earliest examples of HPM applied to irrigation water valuation are by Milliman (1959) and Hartman and Anderson (1962). Daniere (1994) employs this method to investigate urban households' valuation of potable water in Cairo, Egypt. Koundouri et al (2003) apply this method in Cyprus to estimate the effect of water salinity on land prices. Latinopolous et al. (2004) utilise the hedonic pricing method to estimate the implicit value of irrigation water in Chalkidiki, a typical rural area in Greece. Noteworthy applications of this method in developing country context include Gundimeda et al (2003), who value improved water availability and quality in Chennai, India by using the HPM, and more recently Yusuf et al

(2005), who use this method to estimate the WTP for water services using data from the Indonesian housing market.

5.3.2 Travel Cost Method

The travel cost method (TCM) is used to estimate use values associated with ecosystems or sites (such as forests, wetlands, parks, and beaches) that are used for recreation to which people travel for hunting, fishing, hiking, or watching wildlife. The basic premise of the TCM is that the time and travel cost expenses that people incur to visit a site represent the "price" of access to the site. Thus, peoples' WTP to visit the site can be estimated based on the number of trips that they make at different travel costs. This is analogous to estimating peoples' WTP for a marketed good based on the quantity demanded at different prices. The TCM encompasses a variety of models, ranging from the simple single-site TCM to regional and generalised models that incorporate quality indices and account for substitute sites (CGER, 1997).

The method can be used to estimate the economic benefits or costs resulting from changes in access costs for a recreational site, elimination of an existing recreational site, addition of a new recreational site and changes in environmental quality at a recreational site. There are however several limitations to TCM. Defining and measuring the opportunity cost of time is complicated since there is no strong consensus on appropriate measure. Substitute sites are only taken into account in the random utility approach to TCM, which uses information on all possible sites that a visitor might choose, their quality characteristics, and the travel costs to each site. This approach yields information on the value of characteristics in addition to the value of the site as a whole. TCM however can only be used to value goods consumed in situ and, similar to HPM, it cannot capture the non-use values of environmental resources.

The TCM was first proposed by Hotelling (1931) and subsequently developed by Clawson (1959), and Clawson and Knetsch (1966). Such models have been employed to measure the welfare effects to changes in water quality of recreational sites (e.g. Caulkins et al., 1986; Smith and Desvousges, 1986; Bockstael et al., 1987). Noteworthy applications of the method in developing countries include Choe et al. (1996) who apply the TCM to estimate the local community's valuation of surface water quality improvements in the rivers and seawater in Davao, Philippines. Yapping (1998) employs the TCM in China to estimate the value of improving the water quality of East Lake in Wuhan. The results reveal that lake users are WTP significant amounts for the use of the lake and its facilities, thus offsetting some of the cost of maintaining water quality for recreation. Maharana et al (2001) use TCM to estimate the demand curve for visits to the sacred Khecheopalri Lake in India, and then derive the annual consumer surplus accrued to the lake using the number of pilgrims in year 1998.

5.3.3 Other Revealed Preference Methods

In addition to the HPM and the TCM, there are also other revealed preference methods that are not as widely used in the context of environmental resources valuation; however they can be useful in certain situations. These are described below.

Replacement Cost Method. This method values the costs of replacing damaged assets, including environmental assets, by assuming these costs are estimates of the benefit flows from avertive behaviour. This method assumes that the damage is measurable and that the value of the environmental asset is no greater than the replacement cost. It also assumes that there are no secondary benefits arising from the expenditures on environmental protection. This method is particularly applicable where there is a standard that must be met, such as a certain level of water quality (Markandya et al., 2002).

Avertive Expenditures Method. This method is based on the household production function theory of consumer behaviour. The household produces consumption goods using various inputs, some of which are subject to degradation by pollution. In the context of water resources, households may respond to increased degradation of these inputs in various ways that are generally referred to as averting or defensive behaviours so as to avoid the adverse impacts of water contaminants. This includes buying non-durables (e.g., bottled water), making expenditures on liming to reduce water acidification, and changing behaviour to avoid exposure to the contaminant (e.g., boiling water for cooking and drinking or reducing the frequency or length of showers if a volatile organic chemicals were present).

There are however important limitations to this method. Individuals may undertake more than one form of averting behaviour in response to an environmental change and the averting behaviour may have other beneficial effects that are not considered explicitly (e.g., the purchase of bottled water to avoid the risk of consuming polluted supplies may also provide added taste benefits). Furthermore, averting behaviour is often not a continuous decision but a discrete one, e.g. a water filter is either purchased or not. Generally, the averting expenditures does not measure all the costs related to pollution that affect household utility and are therefore only able to provide a lower bound estimate of the true cost of increased pollution.

Applications of the method in Mediterranean countries include Haruvy et al (2000) who apply an optimisation model and develop an economic assessment procedure to asses the costs of averting groundwater pollution, by valuing the damage with the cost needed to treat irrigation water in Israel. Rinaudo et al (2005) use an approach similar to the Avertive Expenditures Method, namely the Avoidance Cost method combined with a Contingent Valuation study, to estimate the cost of groundwater pollution in the upper Rhine Valley in France. Examples of applications of this method in developing countries include those by McConnell and Rosado (2000), who have estimate the non-marginal benefits from improvements in drinking water quality using defensive inputs in Guarapari and Grande Vitoria, Espirito State, Brazil, and Um et al., (2002) who have employed the method to estimate improved drinking water quality in Pusan, Korea.

Production Function Approach. This approach can be used to value non-marketed goods and services that serve as an input to the production of marketed goods. The approach relates the output of particular marketed goods or services (e.g. agricultural production, timber, fish catch) to the inputs necessary to produce them. These include marketed inputs such as labour, capital, and land, as well as non-marketed goods and services such as soil stability, air quality, or water quality and quantity. Thus, the implicit value of water can also be calculated by measuring the contribution of water to the profit in cases where water is an important component of a production process and the producer's cost structure is known. If water supply is unrestricted, a producer will continue to use units of water up to the point where the contribution to profit of the last unit is just equal to its cost to the firm. Even if water is "free", there will be costs to the producer associated with water use (including pumping and delivery costs). If water supply is restricted (for example, by quotas or water rights), the producers may cease use of water before the equality is met. The level of water use at varying costs to the producer defines a "derived" demand relationship, since the demand for the water is derived from the demand for the output of the producer (e.g., agricultural commodities). Yaron (1967) employs a water production function for agricultural products to analyze estimate the demand for water in Israel. In an example from southern Europe, Giannias et al (1996) use this approach to estimate the value of water in a bilateral international context, estimating the WTP of Greece to Bulgaria, for increased downstream water quantity from river Nestos.

Net Factor Income. The Net Factor Income approach estimates changes in producer surplus (i.e., the monetary measure of net benefit to a firm of producing a good) by subtracting the costs of other inputs in production from total revenue, and ascribes the remaining surplus as the value of the environmental input (Brander et al., 2004). Thus for example, the economic benefits of improved water quality can be measured by the increased revenues from greater agricultural productivity when water quality is increased. Alternatively, water quality affects the costs of purifying municipal drinking water hence economic benefits can be measured by the decreased costs of providing clean drinking water.

Cost-of-Illness (COI) method. Another approach is the Cost-of-Illness (COI) method in which the benefits of pollution reduction are measured by estimating the possible savings in direct out-of-pocket expenses resulting from illness (e.g., medicine, doctor and hospital bills) and opportunity costs (e.g., lost earnings associated with the sickness). Two important limitations of this approach is that it does not consider the actual disutility of those who are ill, nor does it account for the defensive or averting expenditures that individuals may have taken to protect themselves (CGER, 1997).

Market Prices. Market prices are used to value the costs/benefits associated with changes in quality and quantity of environmental goods that are traded in perfectly functioning markets. They are generally used with other revealed preference methods (e.g. cost-of-illness approach, replacement costs approach), which assume that market price represents the opportunity cost of water resources. Varela-Ortega et al (1998) estimate the effects of alternative water pricing policies on farmers' water consumption in Spain, while Pujol et al (2006) examine the impact of water markets establishment in Italy and Spain. Market prices and the prices of substitute goods in applications on developing countries are used by Bann (1997) to value the benefits of shifting from the traditional use of the mangrove in Koh Kong Province in Cambodia to commercial shrimp farming. She finds support of retaining existing uses both direct including local fishing and charcoal production and indirect such as storm protection Acharya et al (2002) analyze domestic demand for groundwater in Northern Nigeria with the purpose of valuing the groundwater recharge function of wetlands. They find that the populations in the study area would suffer severe welfare loss if wetlands were to cease providing the existing daily level of groundwater recharge.

5.4 Stated Preference Methods

Stated preference methods (SPM), also called direct valuation methods, have been developed to solve the problem of valuing those environmental resources that are not traded in any market, including surrogate ones. In addition to their ability to estimate use values of any environmental good, the most important feature of these survey-based methods is that they can estimate the non-use values, enabling estimation of each component of TEV. Since many of the outputs, functions and services that water resources generate are not traded in the markets, SPM can be used to determine the value of their economic benefits.

5.4.1 Contingent Valuation Method

The purpose of the contingent valuation method (CVM) is to elicit individuals' preferences, in monetary terms, for changes in the quantity or quality of non-market environmental resources. With CVM, valuation is dependent or 'contingent' upon a hypothetical situation or scenario whereby a sample of the population is interviewed and individuals are asked to state their maximum WTP (or minimum willingness to accept (WTA) compensation) for an increase, or decrease, in the level of environmental quantity or quality. To conduct a CVM, special attention needs to be paid to the design and implementation of the survey. Focus groups, consultations with relevant experts, and pre-testing of the survey are important pre-requisites. Decisions need to be taken regarding how to conduct the interviews (in-person, via mail or via telephone surveys); what the most appropriate payment bid vehicle is (e.g., an increase in annual taxes, a single-one-off payment, a contribution to a conservation fund, among others, see Champs et al. (2002) for more on this); as well as the WTP elicitation format (see Hanemann, 1994; Bateman et al., 2003). Ultimately, the mean WTP bids that have been obtained from the sample can then be extrapolated across the population to obtain the aggregate WTP or value of the environmental resource (Mitchell and Carson, 1989).

With regard to water resource applications, CVM is useful for examining direct use values such as recreational fishing and hunting, and indirect use values such as improved water quality. Unlike revealed preference methods, CVM is also able to measure the option use values of water associated with biodiversity, as well as the non-use values. Despite the strengths of CVM regarding its ability to estimate non-use values and evaluate irreversible changes, this method has been criticised on grounds of lack of validity and reliability (Kahneman and Knetsch, 1992; Diamond and Hausman, 1994). This is on account of potential problems including information bias, design bias (starting point bias and vehicle bias), hypothetical bias, yea-saying bias, strategic bias (free-riding), substitute sites and embedding effects. To address these, the Blue Ribbon Panel under the auspices of U.S. National Oceanic and Atmospheric Administration (NOAA) have made recommendations regarding best practice guidelines for the design and implementation of contingent valuation studies that will form the basis of natural resource damage litigation actions (Arrow et al., 1993).

To date more than 5000 CVM studies have been conducted in over 100 countries, most of which make reference to the guidelines of the NOAA panel, and a large proportion of CVM studies have examined water quality and quantity issues specifically. A number of these have been carried out in the Southern European and Mediterranean countries. There are several notable examples, the earliest of which is by MacPhail. (1994) who employs this method in Tunisia, to estimate households' WTP for piped water and sewer services in Tunis. In the first major application of the method in Italy, Press (1995), estimates public's WTP for improvements in groundwater quality in Milan. Bonnieux et al. (1998) apply the CVM to investigate farmers' willingness to accept (WTA) compensation for provision of wetland conservation in France, as a part of the Environmentally Sensitive Area schemes that are being developed and implemented throughout the European Union, under the obligations of the reformed Common Agricultural Policy. Similarly, Franco et al. (2001) employ a CV in order to assist development of agri-environmental schemes in Italy, by estimating the farmers' valuation of agroforestry and the resultant water quality. Kontogianni et al., (2003) estimate public's WTP to ensure the full operation of a wastewater treatment plant, leading to significant improvements in the water quality of Thermaikos Bay in Greece, with this method. This method is also employed in Turkey, by Ozsabuncuoglu (1996) who estimate farmers' WTP for higher quality of irrigation water in Oguzeli, Gaziantep, and by Goksen et al. (2002) who estimate the citizens' WTP for reduced water pollution in the Bosphorus, Istanbul. In Israel Al-Ghuraiz and Enhassi (2005) estimate the WTP for improved water supply in the Gaza strip. Most recently, Birol et al (forthcominga) employ the CVM to investigate farmers' willingness to accept recharging of Akrotiri aquifer in Cyprus with recycled wastewater, and their WTP for different qualities of recycled wastewater. They find that farmers are WTP the highest for highest quality and quantity recycled wastewater.

The Contingent Valuation Method has been also applied widely to value water resources in developing countries. Whittington et al (1992) employ this method to value water supply options, namely public taps and private connections in Anambra, Nigeria. Choe et al. (1996) compare the results from CVM and TCM to evaluate surface water quality improvements in the rivers and sea-water near the community of Davao, Philippines. Their CV results indicate that household WTP for environmental amenities such as improved water quality is low. Barton (2002) employs this methods to estimate the value of improvements in water quality in Costa Rica. The loss of economic benefits from decreasing quality of water resources was estimated in Vietnam by Phuong (2003). Furthermore, a large number of CVM studies focus on the use and non-use values of wetlands. This is because of the substantial local and global indirect and non-use values inherent in this resource (see Crowards and Turner, 1996; Brouwer et al., 2003 for a review). Of the wetland valuation studies reviewed, a considerable number of them are specific to the Southern European and Mediterranean countries, and several of them are carried out in Greece. Oglethorp and Miliadou (2000) employ the CV method to estimate the use and non-use values of Lake Kerkini in Northern Greece. Kontogianni et al. (2001) employ a CV to evaluate different stakeholders' preferences of four development/ conservation scenarios for the wetland surrounding the Kalloni Bay on the island of Lesvos. The CV study by Psychoudakis et al. (2005) estimates the use values of the several ecological functions of the Zazari–Cheimaditida wetland, including flood water retention, food web support, groundwater recharge, nutrient export and sediment retention. More recently, Birol et al. (2006a) use this method to estimate non-use values the Greek public derives from the sustainable management of the Cheimaditida wetland. CV studies on wetlands from other Southern European and Mediterranean countries include those by Birol et al. (forthcoming b) who estimate the non-use values the Cypriot public derives from the sustainable management of the Akrotiri wetland.

5.4.2 Choice Experiment Method

A relatively new addition to the portfolio of SPM, the choice experiment method (CEM), is theoretically grounded in Lancaster's characteristics theory of value (Lancaster, 1966) and based on random utility models (RUMs) (Luce, 1959; McFadden, 1974). RUMs are discrete choice econometric models, which assume that the respondent has a perfect discrimination capability, whereas the analyst has incomplete information and must therefore take account of uncertainty (see Manski, 1997 for more information). A choice experiment is a highly 'structured method of data generation' (Hanley et al., 1998), relying on carefully designed tasks or "experiments" to reveal the factors that influence choice. The environmental resource is defined in terms of its attributes and levels these attributes would take with and without sustainable management of the resource. For example one attribute that can be used to describe the quality of coastal waters is bathing water quality. The levels of this attribute could be high, medium, and low. One of the attributes is a monetary one, which enables estimation of WTP. Profiles of the resource in terms of its attributes and attribute levels is constructed using experimental design theory, a statistical design theory which combines the level of attributes into different scenarios to be presented to respondents. Two or three alternative profiles are then assembled in choice sets and presented to respondents, who are asked to state their preference (Hanley et al., 1998; Bateman et al., 2003).

Similar to CVM, CEM can estimate economic values for any environmental resource, and can be used to estimate non-use as well as use values. The CEM however, enables estimation not only of the value of the environmental resource as a whole, but also of the implicit value of its attributes, their implied ranking and the value of changing more than one attribute at once (Hanley et al., 1998; Bateman et al., 2003). Another advantage of CEM over CVM is that respondents are more familiar with the choice rather than the payment approach. Moreover, CEM can solve for some of the biases that are present in CVM; the strategic bias is minimised in the CEM since the prices of the resources are already defined in the choice sets. Further, yea-saying bias (or warm glow effect) is also eliminated because the choice approach does not allow for the respondent to state a value for the resource even if they do not value it. Finally, the risk of insensitivity to scope (or embedding effect) in CEM is reduced. If the choice sets offered to respondents are complete and carefully designed, the respondent would not mistake the scale of the resource or its attributes for something else that it could be embedded in (Bateman et al., 2003).

Even though CEM has been applied to valuation of environmental resources only in the past decade, there have been some noteworthy applications of this method to water resources valuation in Southern European and Mediterranean countries in the past few years. Abou-Ali and Carlsson (2004) apply this method to estimate the value of improved water quality in Cairo, Egypt. They investigate the welfare effects of improved health status through increased water quality, and find that the estimated WTP is fairly low compared with the costs of a program that would achieve these improvements. Travisi and Nijkamp (2004) include groundwater contamination from fertilisers and pesticides as an attribute in a survey of willingness to pay for agricultural environmental safety among residents of Milan, Italy. They find that that the public derives substantial economic value from the reduction of groundwater contamination. Colombo et al. (2005; 2006) employ a CE to estimate the benefits from soil conservation measures in the Alto Genil and Guadajoz watersheds in southern Spain. They include surface and ground water quality among the important attributes of the soil conservation measures and find that water quality generated the highest economic value among all the soil conservation measures attributes included in the study.

This method has also been applied to estimate the economic values of several vital components of wetlands Southern European and Mediterranean countries. Nunes et al. (2004) apply the CEM to investigate fishermen's preferences for alternative management practices for clam fishing in a natural wetland, i.e., the Venice lagoon in Italy. Most recently, Birol et al. (2006b) apply the CEM to estimate the Greek public's valuation of the Cheimaditida wetland. They estimate the use and non-use values of several of the wetland attributes, including biodiversity, open water surface area, research and education activities in the wetland and retraining of farmers to environmentally friendly farming practices. They find that the economic benefits that accrues to the public.

Table 5.2 Advantages and disadvantages of economic valuation methods. (Adapted from CGER 1997.)

Method	Advantages	Disadvantages
Hedonic pricing method (HPM)	Based on observable and readily available data from actual behaviour and choices.	Difficulty in detecting small effects of environmental-quality factors on property prices. Connection between implicit prices and value measures is technically complex and sometimes empirically unobtainable. *Ex post* valuation. (i.e., conducted after the change in environmental quality or quantity has occurred) Does not measure non-use values.
Travel cost method (TCM)	Based on observable data from actual behaviour and choices. Relatively inexpensive.	Need for easily observable behaviour. Limited to *in situ* resource use situations including travel. Limited to assessment of the current situation. Possible sample selection problems. *Ex post* valuation. Does not measure non-use values.
Replacement cost method	Based on observable data from actual behaviour and choices. Relatively inexpensive. Provides a lower bound WTP if certain assumptions are met.	Need for easily observable behaviour on averting behaviours or expenditures. Estimates do not capture full losses from environmental degradation. Several key assumptions must be met to obtain reliable estimates. Limited to assessment of current situation. *Ex post* valuation. Does not measure non-use values.
Production function method	Based on observable data from firms using water as an input. Firmly grounded in microeconomic theory. Relatively inexpensive.	Understates WTP. *Ex post* valuation. Does not measure non-use values.
Cost-of-illness method	Relatively inexpensive.	Omits the disutility associated with illness.

(continued)

Table 5.2 (continued)

		Understates WTP because it overlooks averting costs. Limited to assessment of the current situation. *Ex post* valuation.
Market prices	Based on observable data from actual choices in markets or other negotiated exchanges.	Does not provide total values (including non-use values) Limited to assessment of current situation. Potential for market distortions to bias values.
Contingent valuation method (CVM)	It can be used to measure the value of anything without need for observable behaviour (data). It can measure non-use values. Technique is not generally difficult to understand. Enables ex *ante* and *ex post* valuation.	Subject to various biases (e.g., 1. interviewing bias, starting point bias, non-response bias, 2. strategic bias, y 3. ea-saying bias, 4. insensitivity to scope or embedding bias, payment vehicle bias, information bias, hypothetical bias). Expensive due to the need for thorough survey development and pre-testing. Controversial for non-use value applications
Choice Experiment Method (CEM)	It can be used to measure the value of any environmental resource without need for observable behaviour (data), as well as the values of their multiple attributes. It can measure non-use values. Eliminates several biases of CVM Enables ex *ante* and *ex post* valuation.	Technique can be difficult to understand. Expensive due to the need for thorough survey development and pre-testing. Controversial for non-use value applications.

The CEM has also been applied successfully in developing country context. Othman et al. (2004) use this method to assist decision makers in determining the optimal management strategy for the Matang Mangrove Wetlands in Perak State in Malaysia. They estimate the values for environmental attributes (e.g., the area of environmental forest protected, the number of bird species protected and the recreation use of the area) as well as the value of a social attribute (i.e., the employment of local people in wetland based extractive industries). Their results reveal that households experience negative utility from reduced employment and hence demand compensation.

The advantages and disadvantages of the valuation methods described in Sections 5.3 and 5.4 are summarised in Table 5.2 below.

5.5 Conclusions

Values of environmental resources such as water are not straightforward to assess due to the public good nature of this resource. This chapter presents a non-technical introduction to the economic valuation techniques that can be used to capture the total economic value (TEV) of changes in the quantity and quality of environmental resources, with a specific focus on water. Capturing the TEV of water resources is an integral part in the design of economic incentives and institutional arrangements that can ensure their sustainable, efficient and equitable allocation. The chapter also reports the studies that apply valuation methods in Southern European, Mediterranean and Developing countries.

References

Abou-Ali, H. and Carlsson, F. (2004). Evaluating the welfare effects of improved water quality using the choice experiment method, Working Papers in Economics 131, Göteborg University, Department of Economics.

Acharya, G. and Barbier, E. (2002). Using domestic water analysis to value groundwater recharge in the Hadejia-Jama'are Floodplain, Northern Nigeria, *American Journal of Agricultural Economics*, 84(2): 415–426.

Arrow, K., Solow, R., Portney, P.R., Leamer, E.E., and Radner, R.H. (1993). Report of the NOAA Panel on Contingent Valuations. Natural resource damage assessment under the oil pollution Act of 1990. *Federal Register* 58(10): 4601– 4614.

Bann, C. (1997). An economic analysis of alternative mangrove management strategies in Koh Kong Province, Cambodia, economy and environment program for Southeast Asia, International Development Research Centre 1997.

Bateman, I., Carson, R., Day, B., Hanemann, W.M., Hanley, N., Hett, T., Jones-Lee, M., Loomes, G., Mourato, S., Ozdemiroglu, E., Pearce, D.W., Sugden, R., and Swanson, S. (2003). *Guidelines for the Use of Stated Preference Techniques for the Valuation of Preferences for Non-market Goods*. Edward Elgar, Cheltenham,UK.

Barbier, E. (1991). An approach to economic valuation of tropical wetlands: with examples from Guatemala and Nicaragua. In Girvan & Simons (eds), *Caribbean Ecology and Economics*. St. Michael, Barbados: Caribbean Conservation Association.

Barbier, B., Acreman, M., Knowler, D. (1997). Economic Valuation of Wetlands: A guide for policy makers and planners.IUCN.

Barton, D.N. (2002). The transferability of benefit transfer: Contingent valuation of water quality improvements in Costa Rica. *Ecological Economics* 42:147–164.

Birol, E., Karousakis, K., and Koundouri, P. (2006a). Using economic methods and tools to inform water management policies: A survey and critical appraisal of available methods and an application. *Science of the Total Environment*, 365(1–3): 105–122.

Birol, E., Karousakis, K., and Koundouri, P. (2006b). Using a choice experiment to account for preference heterogeneity in wetland attributes: The case of Cheimaditida wetland in Greece *Ecological Economics* 60: 145–156.

Birol, E., Koundouri, P., and Kountouris, Y. Forthcominga. Farmers' demand for recycled water in Cyprus: A contingent valuation approach. In Zaidi, M.K. (Ed.) *Water Reuse: Risk Assessment, Decision-making and Environmental Security*. NATO Science Series, Springer, The Netherlands.

Birol, E., P. Koundouri and Y. Kountouris. Forthcomingb. Water Resource Management and Wetland Conservation. In Koundouri, P. (Ed.) *Water Resource Policy and Related Issues in Cyprus*. RFF PRESS Series, Issues in Water Resource Policy, Resources For The Future, Washington DC, USA.

Bockstael, N.E., Hanemann, W.M., and Kling, C.L. (1987). Estimating the value of water quality improvement in a recreational demand framework. *Water Resources Research* 23(5): 951–960.

Bonnieux, F., Rainelli, P., and Vermersch, D. (1998). Estimating the supply of environmental benefits by agriculture: A French case study. *Environmental and Resource Economics*, 11(2): 135–153.

Boyer, T. and Polasky, S. (2004). Valuing urban wetlands: A review of non-market valuation studies. Department of Applied Economics, University of Minnesota.

Brander, L.M., Florax, R.J.G.M., and Vermaat, J.E. (2006). The empirics of wetland valuation: a comprehensive summary and a meta-analysis of the literature. *Environmental and Resource Economics*, 33(2): 223–250.

Brouwer, R., Langford, I., Bateman, I., and Turner, R.K. (2003). A meta-analysis of wetland ecosystem valuation studies. In Turner, R. K., Jeroen, C.J.M., van den Bergh and Brouwer, R. (eds.) *Managing Wetlands: An ecological economics approach*. Edward Elgar, Cheltenham, UK.

Caulkins, P., Bishop, R., and Bouwes, N., Sr. (1986). The travel cost model for lake recreation: A comparison of two methods for incorporating site quality and substitution effects. *American Journal of Agricultural Economics*, 68(2): 291–97.

Champ, P., Flores, N., Brown, T., and Chivers J. (2002). Contingent valuation and incentives. *Land Economics* 78(4): 591–604.

Choe, K., Whittington, D., Lauria, D.T. (1996). The economic benefits of surface water quality improvements in developing countries: A case study of Davao, Philippines. *Land Economics* 72(4): 519–537.

Chave, P. (2001). *The EU Water Framework Directive: An Introduction*. IWA Publishing.

Clawson, M. (1959). Methods of measuring the demand for and value of outdoor recreation. REF Reprint No. 10. *Resources for the Future*,.Washington, DC.

Clawson, M. and Knetsch, J. (1966). *Economics of Outdoor Recreation*. Johns Hopkins University Press, Baltimore, MD.

Colombo, S., Calatrava, R. J., and Hanley, N. (2006). Analyzing the social benefits of soil conservation measure using stated preference methods *Ecological Economics* 58: 850–861.

Colombo, S., Hanley, N., and Calatrava, R.J., (2005). Designing policy for reducing the off-farm effects of soil erosion using choice experiments *Journal of Agricultural Economics* 56(1): 81–95.

Commission on geosciences, environment and resources (CGER) Valuing ground water: Economic concepts and approaches. http://www.nap.edu/books/0309056403/html/75.html 1997.

Crowards, T.C., and Turner, R.K. (1996). FAEWE sub project report: Economic valuation of wetlands. Centre for Social and Economic Research on the Global Environment. University of East Anglia, Norwich.

Daniere, A.G. (1994). "Estimating willingness-to-pay for housing attributes an application to Cairo and Manila" Regional Science & Urban Economics 24(5): 577–599

Diamond, P.A. and Hausman, J.A. (1994). Contingent valuation: Is some number better than no number? *Journal of Economic Perspectives* 8: 45–64.

Drucker, A.G., Gomez, V., and Anderson, S. (2001). The economic valuation of farm animal genetic resources: a survey of available methods, Ecological Economics, 36:1–18.

Federal guidelines, economic and environmental principles and guidelines for water and related land resources implementation studies. *U.S. Water Resources Council* February 3 1983.

Franco, D., Franco, D., Mannino, I., and Zanetto, G. (2001). The role of agroforestry networks in landscape socioeconomic processes: The potential and limits of the contingent valuation method. *Landscape and Urban Planning* 55(4):239–256.

Giannias, D.A. and Lekakis, J.N. (1996). Fresh surface water resource allocation between Bulgaria and Greece. *Environmental and Resource Economics*, 8, 4:473–483.

Göksen, F., Adaman, F., and Zenginobuz, E.U. (2002). On environmental concern, willingness to pay, and postmaterialist values evidence from Istanbul. *Environment and Behaviour*, 34(5): 616–633.

Griliches Z., ed. (1971). *Price Indexes and Quality Change*. Harvard University Press,Cambridge, MA.

Gundimeda, H. and Kathuria, V. (2003). Estimation of economic value of water scarcity and quality in Chennai, India: The Hedonic approach, South Asian Network of Economic Initiatives Project.

Hanemann, M. (1994). Valuing the environment through contingent valuation. *Journal of Economic Perspectives*. 8(4): 19–43.

Hanley, N. Wright, R.E., and Adamowicz W.L. (1998). Using choice experiments to value the environmental. *Environmental and Resource Economics*, 11(3–4): 413–428.

Hanley, N. and Spash, C. (1995). *Cost Benefit Analysis and the Environment*. Edward Elgar Publishing, England.

Haruvy, N., Hadas, A., Ravina, I., and Shalhevet, S. (2000). Cost assessment of averting ground-water pollution. *Water Science and Technology*, 42, 1–2:135–140.

Hotelling H. (1931). The economics of exhaustible resources. *Journal of Political Economy*. 39: 1937–75.

IUCN, World Conservation Union, (2005) www.iucn.org.

Kahneman D, and Knetsch J.(1992). Valuing public goods: The purchase of moral satisfaction. *Journal of Environmental Economics and Management*, 22:57–70.

Kontogianni, A., Skourtos, M., Langford, I., Bateman, I., and Georgiou, S., (2001). Integrating stakeholder analysis in non-market valuation of environmental assets. *Ecological Economics*, 37: 123–138.

Kontogianni, A., Langford, I.H., Papandreou, A., and Skourtos, M.S. (2003). Social preferences for improving water quality: An economic analysis of benefits from waste water treatment *Water Resources Management*, 17(5): 317–336

Koundouri, P. Three approaches to measuring natural resource scarcity: Theory and application to groundwater. PhD Thesis, Department of Economics, University of Cambridge, UK

Koundouri, P. and Pashardes, P. (2003). Hedonic price analysis and selectivity bias: Water salinity and demand for land. *Environmental and Resource Economics*, 26:45–56

Krutilla, J.V. (1967). Conservation reconsidered, *American Economic Review*, 57(3): 777–786.

Lancaster, K. (1966). A new approach to consumer theory. *Journal of Political Economy*, 1966; 84: 132–157.

Latinopoulos, P., Tziakas, V., and Mallios, Z. (2004). Valuation of irrigation water by the hedonic price method: A case study in Chalkidiki,Greece,*Water, Air, and Soil Pollution: Focus* 4(4–5): 253–262.

Luce, D. (1959). *Individual Choice Behaviour*, John Wiley, New York, NY.

MacPhail, A. (1994). Why don't households connect to the piped water system? Observations from Tunis, Tunisia, *Land Economics* 70(2): 189–196.

Manski C. (1977). The structure of random utility models. *Theory and Decision 1977*; 8: 229–254.

Maharana, I., Rai, S.C., and Sharma, E. (2001). Valuing ecotoursim in a sacred lake of the Sikkim Himalaya, India. *Environmental Conservation*, 27(3):269–277.

Markandya, A. Harou, P., Bellu, L., and Cistulli, V. (2002). *Environmental Economics for Sustainable Growth: A Handbook for Practitioners*, Edward Elgar, Cheltenham, UK.

McConnell K, and Rosado M. (2000). Valuing discrete improvements in drinking water quality through revealed preferences. *Water Resources Research*, 36(6):1575–1582.

McFadden, D. (1974). Conditional logit analysis of qualitative choice behaviour. In Zarembka P. (Ed.) *Frontiers in Econometrics*, Academic Press, New .York.

Milliman, J.W. (1959). Land values as measures of primary irrigation benefits. *Journal of Farm Economics*, 41(2): 234–243.

Mitchell, R.C. and Carson, R.T. (1989). *Using Surveys to Value Public Goods: The Contingent Valuation Method.* Baltimore: Johns Hopkins University Press.Nunes, P.A.L.D., Rossetto, L., and de Blaeij, A., (2004). Measuring the economic value of alternative clam fishing management practices in the Venice Lagoon: results from a conjoint valuation application, *Journal of Marine Systems*, 51:309–320.

Oglethorp, D., and Miliadou, D. (2000). Economic valuation of non-use attributes of a wetland: A case study of Lake Kerkini. *Journal of Environmental Planning and Management*, 43(6): 755–767

Othman, J., Bennett J., and Blamey R.(2004). Environmental values and resource management options: A choice modeling experience in Malaysia. *Environment and Development Economics*, 9: 803–824.

Ozsabuncuoglu, I.H. (1996). Evaluation of negative externality by pollutees *Environment and Planning C: Government and Policy* 14(4): 489–500.

Pearce, D.W. and Turner, K.(1990). Economics of natural resources and the environment. Harvester Wheatsheaf.

Pearce, D.W. (2001). Lecture Notes, University College London, London, UK.

Phuong, D.M. and Gopalakrishnan, C. (2003). An application of the contingent valuation method to estimate the loss of value of water resources due to contamination: The case of the Mekong Delta, Vietnam. *Water Resources Development* 19(14): 617–633,

Portney, P.R. (1994). The contingent valuation debate: Why economists should care. *Journal of Economic Perspectives*, 8: 3–17.

Press, J., (1995). Establishing priorities for Groundwater Quality: A Contingent Valuation Study in Milan, *FEEM (Fonidazione Eni Enrico Mattei)* Newsletter (1): 7–10.

Pujol, J., Raggi, M., and Viaggi, D. (2006). The potential impact of markets for irrigation water in Italy and Spain: A comparison of two study areas. *Australian Journal of Agricultural and Resource Economics*, 50(3):361–380.

Psychoudakis, A., Ragkos, A., and Seferlis, M., (2005). An assessment of wetland management scenarios: The case of Zazari-Cheimaditida (Greece). Water science and technology. *Water Supply Journal*, 5(6): 115–123.

Ramsar Convention. Strategic Plan 1997–2002. Ramsar Convention Bureau, Gland, Switzerland. 1996.

Rinaudo, J.-D., Arnal, C., Blanchin, R., Elsass, P., Meilhac, A., and Loubier, S. (2005). Assessing the cost of groundwater pollution: The case of diffuse agricultural pollution in the Upper Rhine valley aquifer. *Water Science and Technology*, 52, (9):153–162.

Smith, K, and Desvousges, W.H. (1986). *Measuring Water Quality Benefits*. Boston, MA: Kluwer-Nijhoff.

Swanson, T. and Johnston, S. (1999). *Global Environmental Problems and International Environmental Agreements: The Economics of International Institution Building*, Edward Elgar, Cheltenham, UK.

Travisi, C., and Nijkamp, P., (2004). Willingness to pay for agricultural environmental safety: Evidence from a survey of Milan, Italy, Residents Tinbergen Institute Discussion paper TI, 04-070/3.

Um, M., Kwak, S., and Kim, T. (2002). Estimating willingness to pay for improved water quality using averting behavior method with perception measure, *Environmental & Resource Economics* 21: 287–302.

United Nations Environment Program (UNEP) (2000). *Global Environmental Outlook*. Earthscan Publications.

Varela-Ortega, C., Sumpsi, J.M., Garrido, A., Blanco, M., and Iglesias, E. (1998). Water pricing policies, public decision making and farmers' response: Implications for water policy. *Agricultural Economics*, 19(1–2):193–202.

United Nations Environment Program (UNEP) (2005). Vital water statistics. http://www.unep.org/vitalwater/

Woodward, R.T., and Wui, Y.S. (2001). The economic value of wetland services: A meta-analysis. *Ecological Economics*, 37: 257–70.

Whittington, D., Smith, V.K., Okorafor, A., Okore, A., Liu, J.L., and McPhail, A. (1992). Giving Respondents Time to Think in Contingent Valuation Studies: A Developing Country Application, *Journal of Environmental Economics and Management*, 22: 205–225.

Yaron, D., (1967). Empirical analysis of the demand for water by Israeli agriculture. *Journal of Farm Economics*, 49 (2): 461–473.

Yapping, D. (1998). The value of improved water quality: An application of contingent valuation and travel cost methods to East Lake in Wuhan, China, EEPSEA Research Report.

Yusuf, A. and Koundouri, P. (2005). Willingness to pay for water and location bias in hedonic price analysis: Evidence from the Indonesian housing market, *Environment and Development Economics*, 10:821–836.

Chapter 6
The Case for Declining Long-Term Discount Rates in the Evaluation of Flood-Defence Investments

Phoebe Koundouri

There is something awkward about discounting benefits that arise a century hence. For even at a modest discount rate, no investment will look worthwhile.[1]

6.1 Introduction

Debates about discounting have always occupied an important place in environmental policy and economics. Like all other investment, investment in the environment involves incurring costs today for benefits in the future. Whether a public investment is efficient or not is determined by social cost benefit analysis (CBA). In a competitive economy, the socially efficient level of investment is attained by investing in projects where the net present value (NPV), determined by discounting costs and benefits at the social discount rate (SDR) over the time horizon, is greater than zero. It follows that the level of the SDR is critical in determining whether an individual public investment or policy will pass a CBA test.

A common critique of discounting is that it militates against solutions to long-term environmental problems. The question arises: What is the appropriate procedure for such long-time horizons? There is wide agreement that discounting at a constant positive rate in these circumstances is problematic, irrespective of the particular discount rate employed. With a constant rate, the costs and benefits accruing to generations in the distant future appear relatively unimportant in present values terms. Hence decisions made today on the basis of CBA appear to tyrannise future generations and in extreme cases leave them exposed to potentially catastrophic consequences. Such risks can either result from current actions, where future costs carry no weight, e.g., nuclear decommission, or from current inaction, where the future benefits carry no weight, e.g., climate change. The intergenerational issues associated with discounting have puzzled generations of economists. Pigou (1932) referred to the deleterious effects of exponential discounting on future welfare as a 'defective telescopic faculty'. More recently Weitzman (1998) summarises this

[1] *The Economist* (1991), March 23, p 73.

P. Koundouri (ed.), *Coping with Water Deficiency*, 157–163.

puzzle succinctly when he states: 'to think about the distant future in terms of standard discounting is to have an uneasy intuitive feeling that something is wrong, somewhere'.

Discounting also appears to be contrary to the widely supported goal of 'sustainability' which by most definitions implies that policies and investments now must have due regard for the need to secure sustained increases in per capita welfare for future generations (Wald Commission on Environment and Development 1987, Atkinson et al., 1997). Also, by attaching little weight to future welfare conventional discounting appears to ignore any notion of intergenerational equity.

A recently proposed solution to this problem is to use a discount rate, which declines with time, according to some predetermined trajectory, this raising the weight attached to the welfare of future generations. It is immediately obvious that using a declining discount rate (DDR) would make an important contribution towards meeting the goal of sustainable development.

So, what formal justifications exist for using a DDR and what is the optimal trajectory of the decline? This chapter provides a brief non-technical review of recent contributions addressing these two issues in different ways. We tie together the different approaches – some deterministic, others based on uncertainty, some based upon intergenerational equity, others on considerations of efficiency – and we illustrate through a case study on investment in flood defences the effects of using DDRs in public policy. We believe that this work has important implications for the implementation of the EU Water Framework Directive and should be integrated in the economic aspects of such an implementation at the local, regional, country and EU level.

6.2 A Non-Technical Review of the Relevant Theory

In the last decade, the nature of the problem with long-term discounting has become clearer. Four recent theoretical approaches, represented diagrammatically in Fig. 6.1, conclude that this 'awkwardness' can and should be resolved by employing discount rates that decline over time.

First, experimental evidence shows that people discount the future at a declining rate, roughly approximated by a hyperbolic function (Cropper et al., 1999). Results imply discount rates that decline rapidly over the first five to ten years, but start at a surprisingly high level that seems inconsistent with market evidence.

Second, Heal (1998), Chichilnisky (1996, 1997) and Li and Löfgren (2000) present ethical reasons, based upon avoiding a 'dictatorship of the present over the future', which lead to a declining utility discount rate. The main disadvantage of this approach is that it requires estimates of several parameters that would be contentious and possibly also arbitrary.

Third, Gollier (2002) shows that when future consumption growth is uncertain the appropriate discount rate is falling over time. However, Gollier's basic model

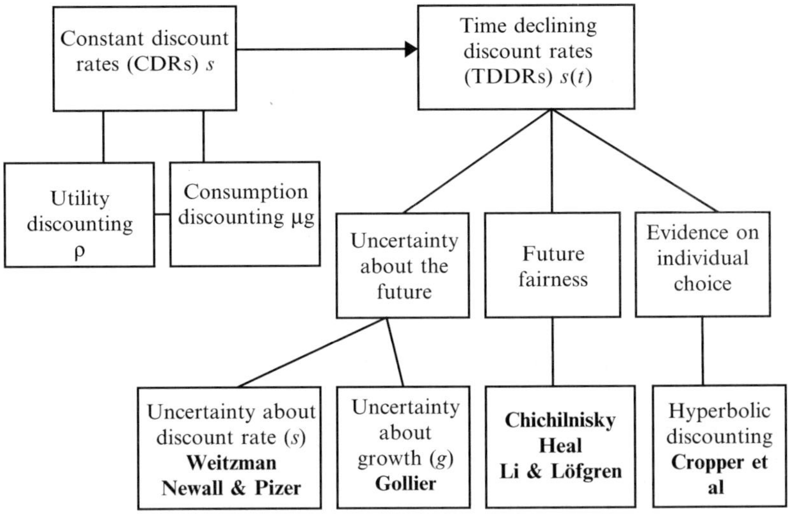

Fig. 6.1 Different discounting approaches

Table 6.1 Proposed step schedule of discount rates

Time from present	Marginal discount rate (%)
1–20 years	3.5
21–75 years	2.0
76–300 years	1.0
More than 300 years	0

assumes a zero risk of recession, a more realistic model leads to considerably more complex results. His approach, while theoretically elegant, proves to be difficult to apply to policy questions.

Finally, Weitzman (1998) shows that any uncertainty in the discount rate leads to a declining discount rate over time. Newell and Pizer (2000, 2001) specify the discount rate uncertainty by running simulations of future interest rates using US interest rate data. Weitzman (2001) specifies the discount rate uncertainty by conducting a survey of over 2000 economists. The responses approximately follow a gamma distribution, leading to a steadily declining discount rate over time. This approach is theoretically sound, simple and of general application.

On balance, it is concluded that the approach in Weitzman (1998) is the most attractive. Moreover, the Weitzman (2001) survey results in a discount rate schedule, shown in Table 6.1 below, consistent with the current Treasury guidelines. As the marginal discount rate is 3.5% for the first 20 years, projects with short time-horizons can be evaluated using conventional methods. Projects with time-horizons over 20 years could employ sequentially lower discount rates.

6.3 Illustration: Flood Defenses

In this section, the discounting theories presented in Section 6.2 are applied to a water-management policy question, with the aim of showing how the use of declining discount rates will affect the appraisal of relatively long-term government policies, programmes, and projects. In particular, this section provides a specific illustration in the area of flood defences investment. The following six discounting regimes are compared:

- A flat discount rate of 6%
- A flat discount rate of 3.5%
- Gamma discounting (with mean discount rate 4% and standard deviation 3%)[2]
- Gamma discounting, step schedule
- Hyperbolic discounting
- Li and Löfgren[3]

Declining discount rates may also have an effect on the economics of flood protection. Over the last ten years, flood-defence investment has been characterised by annual expenditure that has been assumed to offset significant damage—a cost–benefit ratio much greater than unity.

We use a stochastic model by Binne, Black & Veatch designed to assess the costs and benefits of investment in a particular cell (protected area) of flood defences for Shrewsbury for the Environment Agency. The model determines the net benefit of investment by comparing the damage suffered in a 'do nothing' scenario, with damages in the case where 100-year flood defences have been constructed. The benefits can then be compared with the costs of constructing and maintaining the defences.

Source: Shrewsbury FAS project estimates and OXERA calculations.

Employing a 6% discount rate implies that flood defence investment does not pass the cost–benefit analysis. However, a cost–benefit ratio (CBR) of approximately 1.2 is obtained with a 3.5% discount rate, as shown in Fig. 6.2. Furthermore, flood defences are more attractive under all declining rate regimes than under either a 6% or 3.5% fixed-rate regime. The CBR increases by about 17% when the step schedule of discount rates is employed instead of a flat 3.5% rate.

[2] This is a declining discount rate when discount rate uncertainty follows a gamma distribution with mean 4%, and standard deviation 3%. See Weitzman, M. (2001), 'Gamma Discounting', *American Economic Review*, 91(1): 261–271.

[3] This is a weighted average of undiscounted cash flows, and cash flows discounted at 6%. Li, C,-Z. and Löfgren, K.-G. (2000), 'Renewable Resources and Economic Sustainability: A Dynamic Analysis with Heterogeneous Time Preferences', *Journal of Environmental Economics and Management*, **40**, 236–250.

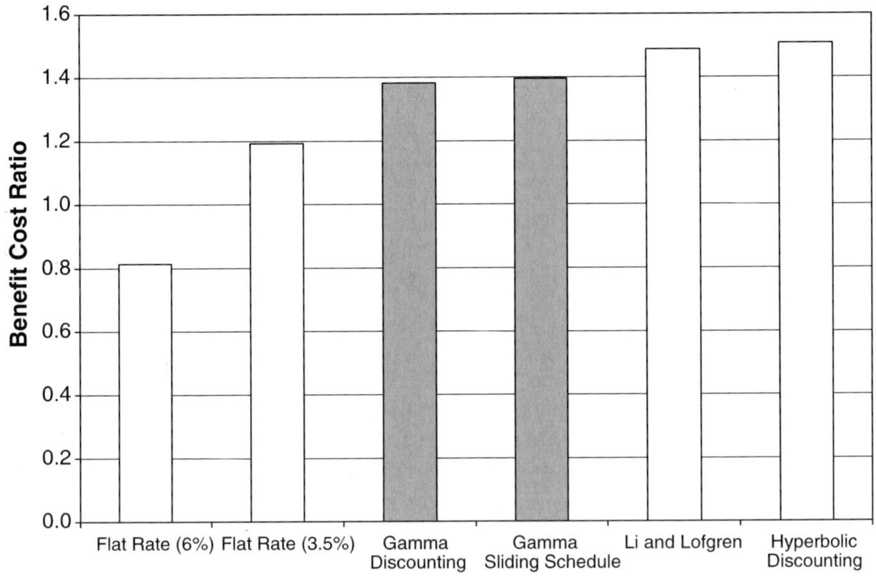

Fig. 6.2 Cost–benefit ratio for a particular cell of flood defences in Shrewsbury

6.4 Policy Implications

The central conclusion from the illustration considered in Section 6.3 and a number of other application of DDR on public policy investment appraisals, namely climate change, biodiversity loss and nuclear waste (see OXERA et al., 2002), is that the impact of declining discount rates depends upon the time-horizon of the project. The longer the time-horizon, the higher the potential magnitude of the effect of declining rates. This dependence is summarised in Table 6.2.

For short-term projects with time-horizons of less than 30 years, declining discount rates have only minimal impact. However, for projects with time-horizons over 30 years, employing declining discount rates may have a significant impact upon the preferred policy. In the road and air examples, shifting from a 3.5% flat rate to the step schedule of rates resulted in an increase in NPV of 8% and 40% respectively.

When time-horizons exceed 100 years, the potential impact is even greater. As Table 6.2 illustrates, it is estimated that the effect could be an increase or decrease of up to approximately ±100% of NPV. For projects with costs and benefits accruing over a 200–400 year time-horizon (such as climate change mitigation), the step schedule of declining discount rates might have an impact of up to approximately ±150% on NPV, relative to discounting at 3.5% constant rate.

Table 6.2 Effect of shift from flat 3.5% to the step schedule of discount rates

Project time horizon	Potential effect on project NPV
0–30 years	Small, generally insignificant
30–100 years	Significant (±50%)
100–200 years	Large impact (±100%)
200–400 years	Major impact (±150%)

6.5 Conclusion

Recent advances in the economic theory of discounting have potentially extremely important implications for policy on energy and on the environment. Whereas the conventional view has always been that there is a unique social discount rate – the value of which has been disputed over thirty years or so of debate – new work suggests powerful reasons why the discount rate is not a single number, but a number that varies in a declining fashion with time. This result emerges from several approaches: from an analysis of how people actually discount the future (hyperbolic discounting); from the implications of uncertainty about the future (the Weitzman and Gollier approaches); and from an explicit attempt to replace the traditional 'present value' maximand of policy appraisal with one that incorporates that goal along with a sustainability requirement. That any one of these approaches could be wrong cannot be doubted, but it seems unlikely that all three arguments can be rejected. Moreover, there is a 'political' argument in favour of the acceptance of time-varying discount rates: in one swoop, they help to resolve the long standing tension between those who believe the distant future matters and those who want to continue discounting the future in the traditional way.

We propose that the conclusions of this chapter and the implications of the theory of DDR, are important for the implementation of the economic aspects, in general, and the investment appraisal, in particular, of the EU Water Framework Directive.

Acknowledgments I am in depth to David Pearce, Ben Groom, Cameron Hepburn and Christian Gollier, for being such an inspiring team while working on the issue of time declining discount rates.

References

Chichilnisky, G., (1996). An axiomatic approach to sustainable development. *Social Choice and Welfare*, 13: 231–257.

Chichilnisky, G., (1997). What is sustainable development? *Land Economics*, 73: 467–491.

Chichilnisky, G., and Heal, G., (1997). Social choice with infinite populations: construction of a rule and impossibility results. *Social Choice and Welfare*, 14(2): 303–319.

Cropper, M., and D. Laibson (1999). The Implications of Hyperbolic Discounting for Project Evaluation. In *Discounting and Intergenerational Equity, Resources for the Future*, pp. 163–172.

Cropper, M. L., Ayded, S.K., and Portney, P.R. (1994). Preferences for Life Saving Programs: How the Public Discounts Time and Age, *Journal of Risk and Uncertainty*, 8: 243–265.

Gollier, C., (2002b). Discounting an uncertain future, *Journal of Public Economics*, 85: 149–166.

Heal, G. (1998). *Valuing the Future: Economic Theory and Sustainability*, Columbia University Press, New York.

Groom, B., Koundouri, P., Panipoulou, K., and Pantelides, T. An Econometric Approach to Estimating Long-Run Discount Rates. *Journal of Applied Econometrics* (forthcoming).

Groom, B., Hepburn, C., Koundouri, P., and Pearce, D. (2005). Discounting the Future: The Long and the Short of it. *Environmental and Resource Economics*, 32: 445–493.

Li, C. Z., and Lofgren, K. G. (2000). Renewable Resources and Economic Sustainability: A Dynamic Analysis with Heterogeneous Time Preferences *Journal of Environmental Economics and Management*, 40: 236–250.

OXERA, Groom, B., Hepburn, C., Koundouri, P., David, P., and Smale, R., (2002). A Social Time Preference Rate for Use in Long-term Discounting. Report to the Office of the Deputy Prime Minister and the Department of the Environment, Food and Rural Affairs. www.odpm.gov.uk/about/discounting/index.html

Newell, R., and Pizer, W. (2000). *Discounting the Distant Future: How Much do Uncertain Rates Increase Valuations?* Discussion Paper 00–45, Resources for the Future, Washington DC. Available at www.rff.org

Newell, R., and Pizer, W. (2003). Discounting the benefits of climate change mitigation: how much do uncertain rates increase valuations? *Journal of Environmental Economics and Management*, 46,1: 52–71.

Pearce, D., Groom, B., Hepburn, C., and Koundouri, P., (2003). Valuing the Future: Recent Advances in Social Discounting. *World Economics*, 4(2): 121–141.

Pigou, A. (1932). *The Economics of Welfare*. 4th edn. Macmillan, London.

Weitzman, M. (1998). Why the far distant future should be discounted at its lowest possible rate, *Journal of Environmental Economics and Management*, 36: 201–208.

Chapter 7
Models and Decisions Support Systems for Participatory Decision Making in Integrated Water Resource Management

Carlo Giupponi and Alessandra Sgobbi

7.1 Introduction

According to the definition provided by the Global Water Partnership (GWP), Integrated Water Resources Management (IWRM) "is a process which promotes the coordinated development and management of water, land and related resources, in order to maximise the resultant economic and social welfare in an equitable manner without compromising the sustainability of vital ecosystems" (GWP-TAC, 2000). The emphasis is thus on the coordinated development and management of water and land resources, with the shared objective of maximising socio-economic welfare in such a way that key ecosystem functions are maintained. IWRM adopts the principles of ecological sciences in terms of system approaches and technical and analytical tools to tackle water management problems. A common paradigm within the context of IWRM is the relevance of the participatory approach, which is becoming a prerequisite of every legislation and plan. According to the GWP again, Public Participation (PP) requires "that stakeholders at all levels of the social structure have an impact on decisions at different levels of water management". Only PP at all levels (international, national regional and local) may assure transparency and accountability of the policy/decision process. In the field of water management, integrated approaches to the resource imply the need for considering the social aspects of water use, as well as the economic and environmental spheres.

In parallel to the widespread adoption of IWRM principles and the increasing emphasis on PP, growing attention is given to the role that could be played by information and communication technologies (ICT) for IWRM, both in shaping policy making processes and in providing an exchange platform to facilitate interactions among different interested parties. Within the broadest category of ICT tools, particular emphasis is placed in this context to simulation models and Decision Support Systems (DSS's). It is beyond the scope of the present work to go deeply in the discussion of either modelling or decision-making theories, nevertheless two considerations are of order. First of all, projections about future developments, or about the impacts of proposed policies, are needed for informing decision making. Simulation models provide the only scientifically based approach to introduce in the process projections and predictions, which are prerequisites for any planning

P. Koundouri (ed.), *Coping with Water Deficiency*, 165–186.

activity. Moreover, decision making requires the integration of different disciplines, priorities and objectives within a unified framework. This integration process is a prerequisite for IWRM which is, by definition, a comprehensive management framework that integrates all the different aspects related to water resources – from the underlying ecological and physical aspects, to the socio-economic values and needs (horizontal integration). DSS tools are specifically designed to support decision-making and policymaking processes by facilitating the integration of different disciplines and perspectives within a common framework. DSS's therefore, given the extreme intrinsic complexity of IWRM, present an evident significant positive potential for improving current practices. By combining the need for advanced ICT tools (which can deal with both scientific evidences and subjective knowledge) and robust participatory approaches (which allow to extend participation in the decision-making process outside a small number of experts – including modellers – and decision makers), *participatory modelling* emerges as a possible solution to problems common to both modelling and PP.

By the term "participatory modelling" we designate a process in which the formulation of a conceptual model and its formalisation are carried out by disciplinary experts with the direct involvement of stakeholders. Various techniques are available in this field, such as creative system thinking and brainstorming, cognitive mapping, causal loop diagrams, etc. The participatory formalisation of the underlying model (i.e., the socio-ecosystem affected by the problem addressed) allows the identification of the main components of the system and their linkages, typically by adopting system analysis formalisation in form of stock (state variables), flows (of energy, matter, information) and causal links (e.g., feedback loops).

Participatory modelling provides the basis of shared knowledge upon which discussions and deliberations about the problem in question can subsequently be established, provided that the techniques adopted to elicit actors' views, and the ability of the facilitator to manage the process, are adequate.

Despite the theoretical potential, experience has shown that the uptake of modelling approaches and computer tools to support decision making and policymaking is very limited in practice (see, for instance, Zapatero, 1996). Traditional modelling techniques have limited impacts on policy making, especially with respect to complex systems such as those involved in natural resource management: these problems – often referred to as "wicked" (Rittel and Webber, 1973) "messy" (Ackoff, 1979) and "hazy" – are difficult to structure, as their understanding depends on the analysts' perspectives. In these cases, the integration of policy actors in the model building process is deemed useful (Geurts and Joldersma, 2001).

Many research efforts have recently began exploring means and ways to tap into the yet unrealised support that models and the modelling process could offer to participatory river basin planning for IWRM. It is within this context that this paper originates, with the aim of building on recent research experiences to offer insights into future research needs in support of participatory planning for integrated water management. More specifically, the purpose of this paper is to illustrate how models, in the broad meaning of the term, could support the integration of political and social dimensions in IWRM.

The remainder of this chapter is organised as follows. In order to put the present research into context, the next section analyses in more detail the role of PP in natural resources management, while the following section explores the use of the terms "model" and "DSS" within the IWRM paradigm. The forth section discusses the specific experience of the MULINO Project ("Multi-sectoral Integrated and Operational Decisional Support System for Sustainable Use of Water Resources at the Catchment Scale") for the implementation of the concepts of IWRM, with specific reference to the EU Water Framework Directive and its implementation process, drawing some general lessons for what concerns public participation. The last section presents some concluding remarks and insights for future research agenda for improving the effectiveness of participatory modelling, with focus on the problems typical of the Mediterranean Region.

7.2 Public Participation and Natural Resources Planning and Management

The rationale underlying public participation in decision making is simple and intuitive: the "public" is more likely to accept a policy when it is consulted beforehand, or when it takes active part in its definition. Public Participation is thus intended as a process to improve decision making, by ensuring that (i) decisions are soundly based on shared knowledge, experiences and scientific evidence, (ii) decisions are influenced by the views and experience of those affected by them, (iii) innovative and creative options are considered, and (iv) the new arrangements are workable, and acceptable to the public. Participatory decision making is thus expected to lead to better decisions and strategies, as well as to reduce conflicts and facilitate enforcement. In this respect, it fosters and supports social learning, useful for developing a common understanding of environmental problems (Walters, 2002). It contributes to enhanced efficiency and effectiveness, promoting democratisation and empowerment of less vocal groups of people. Several authors have recently suggested that participation in decision making is an expression of "normative consideration for the democratic principle" (Stirling, 2006), that is, a "combination of deliberation and equitable involvement of parties" (Renn, 2006) as opposed to the traditional view of representative democracy and command and control management.

The concept of public participation as an important prerequisite for achieving sustainable development clearly emerged in international discussions as early as 1992, during the Rio conference. In particular, Chapter 8 of Agenda 21 (UN, 1993) is devoted to public participation, and identifies "information", "integration" and "participation" as key factors for helping countries to achieve a sustainable development.

In the same year, the "Dublin Statement on Water and Sustainable Development"[1] was adopted at the International Conference on Water and the Environment, stressing

[1] The full text can be found at: http://www.wmo.ch/web/homs/documents/english/icwedece.html

the importance of a holistic, comprehensive, multi-disciplinary approach to water resources. Participants to the Conference called for Integrated Water Resource Management (IWRM), based on four guiding principles covering economic, social, environmental and political issues. Principle 2 of the Dublin Statement states that "water development and management should be based on a participatory approach, involving users, planners and policy-makers at all levels" (ICWE, 1992). Within the context of IWRM, public participation implies that the general public is informed about the importance of water and its management, and that decisions are taken at the lowest possible level, with public consultations and the participation of users.

Since 1992, the concept of PP has been integrated in a variety of international treaties and conventions: the Helsinki Convention (UN, 1992) requires at a minimum information disclosure to the public. The Aarhus Convention (UN, 1998b) further formalises the requirements of pubic participation. At the European level, the EU Water Framework Directive (WFD) specifically calls for participation in river basin planning and management, recognising the potential benefits of this approach.

There is now an extensive literature on the importance of public participation in decision making, especially for environmental matters in relation to sustainable development – see, for instance, Pimbert (2004). In recent years, however, fundamental criticisms to the role of, and means to, public participation are being voiced by both practitioners and researchers. Politicians may have strategic incentives to hamper public participation, as they may not be willing or ready to relinquish part of their decision-making power, or they may fear loosing control over the decisional processes. Public participation itself may become an instrument for control. Without an appropriate and consolidated implementation framework, PP can indeed have unintended negative consequences, such as political co-option, or the legitimisation of decisions favoured by the majority of those involved (Mosse, 1994; Nelson and Wright, 1995; and Cooke and Kothar, 2001). Not enough attention has been devoted to the adequate integration of power relations between elite groups and the less powerful (see, e.g., Hildyard et al., 2001 and Taylor, 2001, as well as Hailey, 2001 and Stirling, 2006), with the real (and documented) risk of public participation used as "a means for top-down planning to be imposed from the bottom-up" (Hildyard et al., 2001, p. 60). Furthermore, who should take part in the process is not clearly defined, nor are there agreed upon mechanisms for the correct identification of the stakeholders to be involved in the process – individuals or groups (Swyngedouw et al., 2002).

Despite the critiques to PP, a widespread emphasis on deliberative decision making remains, including at the national and international legislative levels. Deliberative democracy supporters argue that legitimate lawmaking can only arise from the public deliberation of the citizens: this increasing emphasis on the participatory paradigm is leading to a proliferation of techniques and practices for PP, which may provide significant improvements in the management of natural resources in general, and for IWRM in particular. Yet, although PP is both a right and a practical necessity, its forms, mechanisms and functions need to be carefully shaped (Dalal-Clayton and Bass, 2002).

Approaches derived from the disciplines of system analysis and modelling can help in the systematisation and structuring of stakeholders and experts knowledge, offering methodological and techno-scientific solutions to the increasing need for scientifically sound methodologies. These should guarantee the objective and transparent integration of stakeholders' preferences and inputs in the final decision, through structured approaches, which can be successfully applied under a variety of conditions.

7.3 Setting the Problem into Context: Models, DSS and IWRM

Before entering into the discussion of participatory modelling as a means towards the full implementation of IWRM, an explanation of how models and DSSs are interpreted in this chapter is needed. In fact, policy makers and researchers often have a different understanding of the word "model", as well as different expectations over what a model can(not) do.

Traditionally, modelling in IWRM is seen as an applied method for integrating the various components of a river basin system and their interactions from a wide range of disciplinary perspectives, such as hydrological, economic, agronomic and ecological (Letcher and Bromley, 2005). There is therefore an implicit reference to a mathematical formalisation of reality which, in the context of river basin management planning, relates models mostly to physical and ecological processes, such as nutrient balance, sediment transportation, and hydrological models. The social dimension is often not included, as it is difficult to formalise. In this literal sense of the term, "physical" models could be understood as computer tools, in which systems of differential equations are used mainly for research purposes, or for management issues and, sometimes, for forecasting or exploring different future scenarios in the fields of hydraulics, hydrology, geomorphology, water ecology and chemistry; such tools traditionally lack the integration of socio-economic aspects (Hare, 2004b). Designed to answer specific questions in specific settings (e.g., resource allocation, water quality, etc.), such models cannot be easily reapplied in different contexts (McIntosh et al., 2004).

This narrow definition of models as tools has generated some confusion when exploring participation and modelling. As a response, in recent times the role of modelling in the application of IWRM and the definition of "models" has become more encompassing, intended "broadly, to consider both what they represent in the river basin (e.g., run-off, population change, stakeholder perceptions) and how they may be packaged for use" (Hare, 2004b, pp. 43–44). Many experts and practitioners alike understand the word in a broader sense, not limited to physical, mathematical representations of reality, but encompassing mental models and the related cognitive maps as well – thus the social dimension of water management. Mental models refer to an "internal", subjective representation of reality, while mental maps are to be understood in this context as the processes and techniques used for "external" representations of mental models (Doyle and Ford, 1998) – that is, as means to share the subjective view of reality. External representations of mental maps can then be transferred to a wide range of computer tools – either in a formal, mathematical format, providing insights or forecasts in physical processes, as a way of representing and organising knowledge about the (management) system, or of supporting communication and decision making, and, therefore, PP. The term "external" is often omitted, yet it plays a fundamental role for our purposes. In fact, a crucial assumption for effective participatory modelling is that any manifestation of "external" models or modelling activity may exist as representation of (internal) mental models. Cognitive mapping techniques play a key role in ensuring that the emerging external model(s) is a fair enough representation of internal structures and beliefs, as well as a good enough compromise view of the problem under discussion. Vennix et al (1990) offer an extensive review of literature on cognitive mapping, and guidelines on how to structure the knowledge elicitation process. Other methods for problem structuring are discussed in Mingers and Rosenhead (2004), Rosenhead (1989), Rosenhead and Mingers (2001), Flood and Jackson (1991), and Dyson and O'Brien (1998).

Where modelling is applied in decision-making and planning processes, a specific category of broadly indented models come into play: DSS's. Decision Support Systems are commonly defined as tools to support the structuring of a decision-making problem, as well as improving the effectiveness and acceptance of the final policy choice. Referring specifically to natural resource management applications, DSS's can support the organisation of information and knowledge in such a way that policy makers are able to analyse and compare different management strategies, and to integrate their own priorities and value judgments in the decision-making process in a transparent way (Mysiak, 2005).

Only by accepting the challenge of approaching the internal component(s) of models in a participatory context to construct their external counterparts, can we expect to harness the full potentials of modelling for improved resource management. Unfortunately, this has rarely been the case in the recent history of PP. To complicate matters further, the potential role of modelling itself has been questioned in recent times, with decision makers often viewing models (including DSS) as "black boxes" which cannot be fully trusted. The debate about climate change has remarkably contributed to the crisis of credibility of models (van der Sluijs, 2002). There is in fact a general perception that modelling, and in particular scenario models for future projections, remains an academic exercise with very strong components of subjectivity and uncertainties. As such, the results of models cannot be fully trusted, as they could be subject to manipulation by experts, policy makers, or interested groups. Perspectives for the solution of such problems are offered by post-normal science (Funtowicz and Ravetz, 1993), which recognises that scientific and technical discourse should be opened to non-experts (stakeholders and the general public). Only accepting the challenge of opening these "black boxes" by involving the interested actors in the conceptual formulation of the tools can we expect concrete potentials for their future use and, in particular, for the process of implementing the principles of IWRM in the real world.

Over the past decades, quite often DSS's were seen as computer shells to provide user interface to models, and potential end-users were not involved in the development or testing phases. With the increasing importance of public participation in decision making, efforts were made by researchers to involve the potential end-users of the software in the development process, but involvement was still limited in practice. Recognising the shortcomings of the earlier approaches to DSS development, researchers are now increasingly including end-users and stakeholders' input since the early stages of design.

A notable shift has thus occurred, with emphasis moving away from the coding of DSS *tools* to the *process* of developing these tools through participatory approaches. Participatory modelling, therefore, becomes a crucial component of DSS development, with particular attention to ways in which DSS's could be of use for communicating during the decisions process and/or helping finding a compromise solution with the participation of the interested actors. Participatory modelling thus entails the active and direct involvement of stakeholders in model formulation, helping identifying the components of the model and providing inputs to it.

This approach to decision making, combining "soft" and "hard" sciences, is receiving increasing support (Mendoza and Prabhu, 2005; Richards et al., 1995), as the tools for river basin planning may be of greater use to IWRM when developed and applied through a participatory process. This holistic interpretation of "modelling as a process" strengthens the need for participatory modelling, and DSS's can be seen as suitable approaches for providing inputs to facilitate and promote it. DSS's should however be considered in their more innovative and comprehensive definition, no longer as tools aimed at providing "the correct answer" or at identifying the best option among a set of alternatives, but rather as offering methodologies and strategies to support and manage the decisional process in all its stages.

Besides its evident potential, participatory modelling still presents many challenges to both the scientific and the policy communities. To the scientists, it requires the development of novel methodologies and tools for model design and implementation in a participatory context. Policy makers need to invest significant resources for facilitating the development of the new models, and for participating in the process itself. It is however our belief that such efforts are worth for both communities, given the current situation of limited exploitation of research outputs and the clear need for improving the quality of policy and decision making, as required by the most recent evolution of policies and legislation.

7.4 The Mulino Experience with Modelling and Decision Support for IWRM

7.4.1 Water Resources and Policy Issues in the Mediterranean Region

The Mediterranean basin is characterised by a strong heterogeneity of cultures, economies, and societies. This diversity – which is both north-south and across countries on the same shore – has been recognised as a key contributing factor to the present economic marginalisation of the southern bank: only three of the 12 Mediterranean partner countries in the Southern bank have income levels similar to those of some EU countries, and they account for almost ¼ of the region's GDP. This marked heterogeneity has also often implied problematic interactions between Mediterranean countries, both bilateral and multilateral, and to instability in the region, problem which is more acute for the management of transboundary issues, such as trade, pollution, and water.

The environment in the Mediterranean basin is characterised by low resilience, aggravated by the insufficient attention paid to environmental issues during fast industrialisation, both in the Southern and Northern sides of the basin. Moreover, the political and institutional capacity for environmental policy remains low, especially in the South.

The key environmental challenges for the region are water management, waste management, polluted areas and loss of biodiversity, preservation of coastal areas, and desertification. Water is perhaps the most important shared natural resource in the Mediterranean basin. The region is characterised by water scarcity, recurrent droughts and impending desertification, which threaten not only the economic viability of the region, but also its political stability. This situation is expected to deteriorate in the future as a consequence of global climate change and increasing anthropogenic interference with the ecosystems. An aspect of water scarcity, which is too often disregarded, is its link with regional security and stability. Conflicts over access to, and control of, water resources are not new to the region – and can only become more important if no joint management decision is taken.

Water resources are scarce in the Mediterranean area, and yet the current management regimes are at times neither efficient nor sustainable. The need to improve on the current system is paramount, if the objective of sustainable development subscribed by partner countries and the EU and stability in the region are to be achieved and maintained. Within this context the EU launched the *EU Water Initiative (EUWI): Water for Life*, with a specific Mediterranean Component (MED EUWI) aiming to:

- Assist design of better, demand driven and output oriented water related programmes
- Facilitate better coordination of water programmes and projects, targeting more effective
- Use of existing funds and mobilisation of new financial resources and
- Enhanced cooperation for project's proper implementation, based on peer review and strategic assessment

The main reference for the EU approach to water policy and management is the Water Framework Directive (EC/2000/60), whose principles are brought also to international cooperation efforts such as the MED EUWI. The EU WFD embraces the paradigm of IWRM, and it strongly encourages participatory decision making for water resources management.

7.4.2 Public Participation in the Water Framework Directive

A relatively large, albeit recent, body of literature has developed on the role of public participation in the implementation of the WFD, and the Directive's requirements with respect to this aspect (see in particular EC, 2003a).

Public participation in the EU Water Framework Directive (WFD) context can be interpreted as allowing people to influence outcomes and planning processes. The WFD identifies five main benefits of participatory planning: (i) increased awareness of environmental issues; (ii) facilitation of integrating the knowledge, experience and initiatives of different stakeholders in the decision process, thus improving it; (iii) increased probability of public acceptance, commitment and support; (iv) a more transparent and creative decision making; (v) and, finally, the reduced likelihood of confrontations, conflicts, and litigations arising from the decisions and/or its implementation (EC, 2003a).

Specifically, the preamble of Article 14 of the WFD clearly defines the importance of public participation for water management in Europe, and spells out the different level of public involvement for integrated water management. A more detailed and general description of the types and degrees of public participation can be found, for instance, in Arnstein (1969), Allen et al. (2002), Hare et al. (2003).

Art. 14 of the Directive also states that a pre-condition for effective PP is access to information, which shall be ensured. The lowest possible level of PP, which shall also be ensured, is consultation – that is, the public can react to plans and proposals developed by the authorities through the provision of written comments, participation in public hearings, or in surveys and questionnaires. Finally, at the higher level, stakeholders can be involved in all aspects of the implementation of the Directive, especially in the planning process. That is, they can take active part in the process by discussing issues and contributing to their solution. Art. 14 of the WFD urges member states to encourage the active involvement of stakeholders.

7.4.3 Progress with IWRM and PP in Mediterranean countries

Despite the PP requirement, European decision makers seem to find it difficult to comply with this aspect of the WFD. This slow engagement with PP could be due to the belief, on behalf of policy makers, that PP entails relinquishing part of their decision-making power, or to the lack of a widespread "culture of active participation". But it seems more plausible that significant obstacles are to be found in the constraints faced by policy makers. Although there are no extensive surveys on the issue, anecdotal evidences and specific experiences support the hypothesis that policy makers still lack the tools and methodologies for effectively engaging stakeholders in decision-making, as also highlighted by recent research efforts, such as the Harmoni-CA project's activities and workshops (http://www.harmoni-ca.info/), and the surveys conducted within the framework of the NOSTRUM-DSS project (http://www.feem-web.it/nostrum/). For instance, the institutional structure for water governance is and remains strongly centralised in most Mediterranean countries – especially in the Southern bank.

Despite the wide acceptance of the benefits of public involvement in decision-making processes, one can distinguish four main approaches to decision making for water governance in Mediterranean countries (Sgobbi and Frafiga, 2006):

- Top-down approach, where only government institutions and primary actors take part in the decision making. Primary actors are those who are affected by the decision directly. This is the case, for instance, of Tunisia and Syria, and Morocco.
- Top-down approach with the theoretical possibility for secondary actors to participate in the process, but because of various constraints this possibility is not exploited. Intermediaries in the process of decision making are considered as secondary actors. The constraints to their participation can either emerge from existing conflicts among different governance level and individual actors, as is the case for Italy and Lebanon; or due to a lack of communication and dissemination of information, as in the case of Portugal and Turkey.
- Interactive processes, whereby key actors take active part in decision-making activities, and a compromise solution is sought. Key actors are those who can significantly influence the process, or are important in determining the success or failure of an action. Algeria, Cyprus, Spain and Egypt normally follow this route. And
- Bottom-up approaches, where secondary actors and end-users are actively involved and participate both formally and informally, as in Croatia and Greece.

In summary, the participation of actors in water planning and management is still low in several countries, where water resources are highly centralised – such as in Israel and Algeria, but also Lebanon, where a network of water users at the community level is still lacking. In EU countries, public participations should be contemplated in the decision-making processes, but the mechanisms available are often highly inefficient, or means to integrate the preferences of actors in actual policies are not clearly established, with a resulting lack of transparency. A common problem encountered in practice is the reluctance of private actors and government authorities to participate in joint decision-making processes – an obstacle whose roots are probably to be found in the specific social, political and economic situation of individual countries, as well as in the long history of social conflict over water in many countries, with the consequence loss of trust in the other parties.

7.4.4 The Contribution of the Mulino Project to Participatory Modelling

MULINO was part of the EU's efforts to develop operationally useful DSS for IWRM, and was set within the context of the EU water policy, as defined by the WFD. The project – which began in 2001 and was concluded at the beginning of 2004 – had the main objective of addressing the existing gap between researches and policy makers

Table 7.1 Actors in the WFD

(General) public	One or more natural or legal persons, and, in accordance with national legislation or practice, their associations, organisations or groups" (SEIA Directive (2001/42/EC). Aarthus Convention, Art, 2(4)).
Interested parties / Stakeholders	Any person, group o organisation with an interest or "stake" in an issue, either because they will be directly affected or because they may have some influence on its outcome. "Interested parties" also includes members of the public who are not yet aware that they will be affected (in practice most individual citizens and many small NGOs and companies).
Broad public	Members of the public with only a limited interest in the issue concerned and limited influence on its outcome. Collectively, their interest and influence may be significant.

(Source: EC, 2003a)

with respect to the development and use of DSS tools for IWRM, through the design and implementation of an operational decision support system for the management of water resources (Giupponi, 2006). The tool could be based on hydrological modelling, multi-sectoral indicators, and a multi-disciplinary criteria evaluation process, and was intended to provide a framework to integrate quantitative with more qualitative information. The methodology was expected to contribute to the quality and transparency of decision making for the development of River Basin Management Plans, whenever a solution has to be selected within a discrete set of alternative options. The active involvement of actors – both end-users and stakeholders – was sought from the early stages of DSS development, with the aim of exploring ways in which DSS processes could offer support to decision makers for participatory planning.

The main tangible result of the project is a stand-alone DSS software, *m*DSS, which is freely available from the project web site[2], and provides functionalities to support the integration of socio-economic and environmental modelling techniques with GIS functions and multiple criteria decision methods.

When exploring participatory modelling as an indirect means to comply with the WFD requirements of participatory planning, the most relevant aspects of the MULINO Project are the methodology and process within which *m*DSS was developed. In an attempt to bridge the observed gap between DSS development and application, a two-pronged strategy was used: (i) first, a participatory development approach was taken right at the outset through the involvement of potential end-users in the tool development phase; (ii) secondly, the methodology and software were developed for application under a variety of conditions and different spatial scales. The test in 5 countries and in one pan-European application demonstrated the flexibility of both the process and the resulting tool (see Table 7.2).

[2] The latest version of the mDSS software can be downloaded from http://www.netsymod.eu/mdss/ Additional documentation on the project and the software can be found at http://siti.feem.it/mulino/

Table 7.2 The MULINO case studies and their outcomes in brief according to the conclusive notes of researchers and end users involved (from the MULINO project Final Report)

Romania	*Bahlui*	1950 km^2	National	"What is the best farming strategy to minimise sediment and nitrate loads while preserving living standards of rural communities?"
• The use of *m*DSS allowed the collaboration of stakeholders in a way that would not have occurred otherwise. • The tool may be used as a framework for the implementation of the Nitrate Directive in vulnerable zones. • It was difficult to explain to the stakeholders and end-users (even for the Romanian MULINO staff) the link between the DPSIR components and MCA. • It was difficult to explain the different result using TOPSIS.				
Portugal	*Caia*	780 km^2	National	"What is the optimum level of water retention (control) in the Caia dam for multi-sectoral water management?"
• Care should be taken in the involvement of stakeholders to avoid dominating personalities to orient the final decision. • Great care is needed in managing the value functions and weights.				
UK	*Yure & Bare*	2500 km^2	National	"What are the optimal seasonal water prices for maximising irrigation while minimising the adverse ecological impacts?"
• Great care should be taken in managing the relationships between value functions and weights. • The sustainability chart does not express effectively the relationship between the balancing of the 3 pillars and the scores of the options.				
Belgium	*Nethan*	55 km^2	Regional	"How can we reduce the risk of flooding? How large should be storm basins, and where should they be located?"
• The application to past decision making may promote both collaboration between scientists and managers as well as the communication between end-users and stakeholders.				
Italy	*Vela*	100 km^2	Local	"What are the best solutions to reduce the nitrate discharges to the Venice Lagoon from the rivers of the Vela sub-basin?"
• The case study is a very good example of the usefulness of hierarchical weighting and shows the power of mDSS in clearly identifying which options are competitive. It also showed that to be effective the group decision-making procedures need more than two decision makers.				
Italy	*Cavallino*	23 km^2	Local	"How can we substitute groundwater with surface water for irrigation? Which is the best treatment method for guaranteeing water quality standards?"
• mDSS has potential to formalise and standardise the decision process, thus allowing parallel, coherent and comparable, decision processes to be carried out by different stakeholders. It facilitates the identification of dominated options. • It was found an ideal tool to implement an Environmental Impact Assessment and a useful tool for working at the interface between technical and political "bodies" within the administration and support the political side.				

(continued)

Table 7.2 (continued)

• When confronting a wide range of stakeholders, hierarchical weighting is to be preferred to allow for concise and non technical communication, to non experienced people. • mDSS requires a trained end user in the competent administration with time available to invest in learning how to use it. • the description of DPS chains may be a useful support for clarifying the decision background and its interpretation by the decision makers, but its identification is often subjective or ambiguous. • The group decision-making functionalities of mDSS could not solve the conflict found between the different views (i.e., criterion weights) of the stakeholders. This function should be further developed.				
Italy	*Arborea*	100 km^2	Regional	"What is the best way to reduce the contaminants entering the phreatic aquifer in Arborea?"
• The study has shown that a multidisciplinary approach must be adopted to have a clear view of the problem at hand. • The use of physically based models at the catchment scale can be important to predict the effect of different land exploitation schemes on the water quality of downstream water bodies. Still, further evaluation is needed to identify indicators that reflect critical ecosystem processes or state variables related to the integrity and sustainability of those ecosystems. • The DPSIR approach appears to be useful mainly in the conceptualisation and standardisation of the problem at hand. • The mDSS is an appropriate tool to couple socio-economic and environmental indicators, supporting policy makers in exploring the decision context.				
Europe	–	3216000 km^2	European	"What is the most efficient option for spatial implementation of the Nitrate Directive?"
• The indicators and their spatial distribution were calculated by means of external GIS software due to the limited capabilities of the Spatial View function of mDSS, which, nevertheless, facilitates data displaying to the actors involved. • In this case of spatial decision-making, where indicators are represented as maps, single averaged values were extracted in order to fill the analysis matrix, functions to express spatial variability should be considered in future applications.				

The main components of the MULINO approach – which takes experts, end-users and selected stakeholders through the steps of building the decision process with the support of *m*DSS – are illustrated in Fig. 7.1 (for a more detailed description of the methodology and the software, see Giupponi et al., 2004 and Mysiak et al., 2005). Prior to entering into the process, a Social Network Analysis is undertaken, with the purpose of identifying the key stakeholders to involve in the modelling exercise, as well as their relational ties. In the application of MULINO, a methodological approach for the analysis of local networks in the context of IWRM is provided, and applied to the decisional context in the case study. The methodology serves first of all to identify the role played by local actors, but also to involve them in the implementation of the project, and establishing local network of stakeholders in the case studies of reference. A comparison across the MULINO applications highlights how the dimension and density of the network are correlated, and how, in larger network, actors' centrality (and thus their power) seems to be more

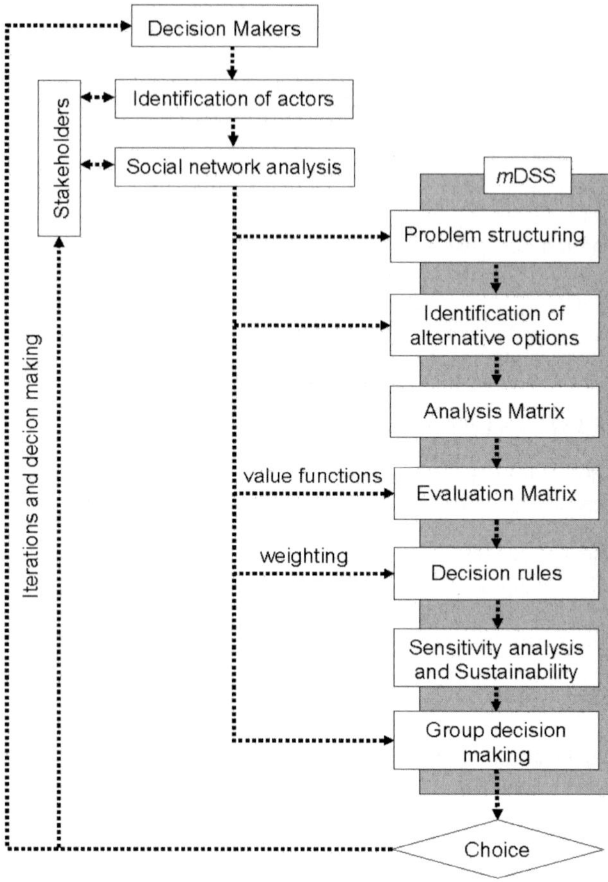

Fig. 7.1 Participatory approach developed by the MULINO Project (from Giupponi, 2006)

unevenly distributed. This consideration is important when establishing water management policies, but also and foremost when planning for public participation (for additional details, see Lourenço et al., 2004 and Feás et al., 2004).

After the preliminary identification and involvement of stakeholders, in the *Conceptual Phase*, (i) the exploration of the decision problem, (ii) the identification of alternative options to address it, and (iii) the selection of key indicators take place, with the active participation of both decision makers and selected stakeholders. The Conceptual Phase represents the MULINO attempt to implement participatory modelling in decision making, when the problem to be addressed is "modelled" with the help of key indicators and the sketching of causal ties. At this stage, models are represented by mental maps which provide the structure of the problems to be addressed, and the relational ties linking each factor and issue. To facilitate and structure participatory modelling, the process is based upon an intuitive and easy-to-grasp approach, the Driving Force-Pressure-State-Impact-Response (DPSIR) framework (EEA, 1999). This approach produces a formal description of activities and issues relevant to the management of water resources, making explicit the causal chains relating human activities and the state of the environment. The DPSIR framework, therefore, supports the process of participation though mental modelling, implementing a system analysis approach, answering to the users' need for more structured and formalised external models. This is achieved through the capacity of the DPSIR framework to simplify the understanding of the complex relationships between the drivers of environmental problems, their impacts, and society's responses to them. This schematisation highlights the causal links between interacting components of social, economic and environmental issues relevant for the management of resources. The DPSIR methodology is a powerful tool for participatory modelling, offering a simple structure to share individual cognitive maps of a system, and facilitating the process of arriving at a shared view of the problem. The *Conceptual Phase* then provides the ground for implementing physical and socio-economic mathematical models in the subsequent *Design phase*.

During the *Design Phase*, the set of feasible alternatives are defined, corresponding to the responses within the DPSIR framework. End-users also identify the set of evaluation criteria for assessing the performance of each option, on the basis of the available information. Hydrological models can be implemented in mDSS through a generic interface, which supports the coherent management of Driver, Pressure and State indicators as catchment variables which are distributed in space and time. The values of the selected indicators for each option are stored in the Analysis Matrix (AM).

In the final phase, the *Choice Phase*, the normalisation and weighting of the multidimensional data stored in the analysis matrix take place, leading to the evaluation of the options that are thus compared and ranked. The selection of the final choice is based on the application of Multi-Criteria Decision Methods (MCDMs), which allows the comparison of alternatives and their performances with respect to predefined and agreed evaluation criteria, and their relative importance (i.e., weights). The *m*DSS tool provides for more than one decision rule to aid the process, and includes the possibility of parallel implementations of the *Choice Phase* to allow for group decision making in a participatory context: in this case, actors may individually or in group apply their own valuing and weighting decisional criteria, which are then combined in a final stage where group decision-making routines are implemented. All the steps of the decision process are stored in documentation files allowing for transparent communication of the subjective judgements and preferences expressed by the users and combination of multiple parallel processes in a multi-stakeholder context.

7.4.5 Lessons Learned and Further Research Developments

From the methodological and empirical applications of DSS's, it emerges that the process of model development has assumed a central role, as opposed to the model itself, either as a pre-existing tool, or as the end product of the process. As such, it is participatory modelling which may more directly and immediately support policy makers in implementing participatory planning in the context of IWRM in general, and the Water Framework Directive in particular.

As illustrated in the preceding sections of the paper, the process of participatory modelling is central to the MULINO methodology, and it is based on the use of a rather intuitive and easy to communicate framework, DPSIR. Despite the notable progress, however, the management of stakeholders' involvement is still poorly organised and supported, and so is the selection of the stakeholders to take part in the process (see, for instance, the case studies reported in Hare et al., 2006). Furthermore, research developments after the end of the MULINO project were focused on improving the software through updating and new developments, but the greatest emphasis was placed on the structuring of the decision process, the provision of methodological solutions to the identified shortcomings of current practices, and the identification of possible tools to support participatory modelling within the newly developed NetSyMoD methodological framework (Giupponi et al., 2005).

There are various strategies for improving participatory modelling, such as providing end-users methodological support for the selection of key stakeholders through the implementation of network analysis techniques, and using cognitive mapping techniques for eliciting stakeholders' views of the problem, in combination with the DPSIR conceptual framework. Social Network Analysis (see, e.g., Wasserman and Faust, 1994) can fruitfully be used for the identification of key stakeholders and their perspectives, as well as for mapping power relations within the network of interested parties. The results can easily be integrated within the design of the DSS tool. This would ensure that the selection of actors to take an active part in the process is carried out in an objective manner, that no key stakeholders are left out, that those taking part are truly representative, and that power relations are managed appropriately in the group participatory exercise, so that all key stakeholders can meaningfully participate. Participatory workshops and elicitation techniques can provide an objective and reliable way to elicit mental maps from the stakeholders, therefore responding to the explicit need of the policy makers' community of moving from haphazard stakeholders' participation to structured participation over long time periods (Hare, 2004a, and Hare, 2005). Various cognitive mapping techniques are available for this purpose, such as the hexagon method (Hodgson, 1992) and causal modelling (Vennix, 1996).

Despite the early stages, participatory modelling seems to offer a good methodological approach to participatory planning and management, and further efforts are likely to be well invested in this field. Nonetheless, some open questions require adequate consideration:

- What incentives can be provided to policy makers to ensure that it is worth for them investing in acquiring knowledge of participatory modelling techniques and/or DSS tools?
- As building DSS tools is a resource-intensive exercise, is it possible to develop a tool based on participatory modelling processes, which is representative of many different situations and can, therefore, be safely used in different circumstances?

- How can the training requirements of potential end-users be best addressed?
- Which avenues do researchers have to develop new methodologies and tools with a higher probability of finding an application in decision-making exercises? Should the DSSs tools be simplified as much as possible to favour their use, or should their technicalities and complexities be maintained?

These remarks may not be entirely new, as these questions have been discussed in the literature for quite some times (see, e.g. Chapters 7 and 8 of Giupponi et al., 2006, and the references therein). Yet, it is also clear that empirical applications often do not give appropriate considerations to the problems so often identified in the literature. It is thus worth reiterating that finding suitable ways of addressing these concerns would significantly contribute to the improvement of public participation in IWRM, as will be discussed in more details in the remainder of this section.

As highlighted by the Harmoni-CA Synthesis Workshop (Hare, 2005), "the lack of use of models by policy makers and the lack of participation in management both have the same root cause: the lack of incentives to do so.": triggering the process of model and DSS adoption by potential end users remains critical. We propose here two possible ways in which policy makers could be motivated to invest in learning how to use DSS development process and the resulting DSS tool:

- A *regulatory* approach, by which using specific participatory modelling and tools becomes a legal requirement, embedded in national and/or European policies
- A *demand driven* approach, in which policy makers are provided with specific incentives to adopt participatory modelling and tools in decision making. A necessary condition for this approach to work is an increased awareness among policy makers of the medium to long-term *benefits* of PP (especially in terms of reduced conflicts and enforcement costs). Furthermore, the demand driven approach can only be effective if the developers of DSS tools and participatory methodologies are able to address the explicit needs of policy makers, and develop tools and framework which respond to the exact legislative requirements and reporting rules, and would thus facilitate compliance.

Early involvement of end-users and stakeholders is often advocated for as a means to ensure that the process of participatory modelling is effective, that the resulting tools are used in actual decisional processes and, therefore, address the needs and requirements of decision makers. Yet, there is a trade-off between the extent of early involvement and the extent to which the tool can be re-used. Indeed, as far as models – and DSS's – are equated to computer tools, the issues of usability and re-usability become crucial (McIntosh et al., 2004). Modularisation of tools within a coherent and general methodological framework may represent an effective strategy to reduce the time and effort needed for building DSS's, without however compromising their tailoring to specific situations.

Even when policy makers are willing to adopt participatory modelling for decision making, adequate capacity building and training of potential end-users remain necessary to ensure the process is not mismanaged, or the tool misused. One of the major barriers to policy makers using models is the complexity of models themselves, which entail a significant investment of time and human resources to understand and master. For the effective use of models, policy makers are thus required a thorough knowledge of elicitation techniques and software use. Three main strategies are suggested to improve capacity building:

- Strengthen the role of trained professionals acting as facilitators in the participatory planning process. Such a professional background should find more attention in the curricula offered by European universities
- Promote an extended campaign of in-house expertise development, enabling policy makers to meaningful undertake participatory planning, and using modelling tools for decision making. For this strategy to be effective, however, one would need to find a "normative" entry point to provide the right incentives, as discussed above
- Increase research efforts in developing tools and methods for the use of consultants, who would then provide external support to decision makers. Before investing substantial resources along this path, however, careful consideration should be given to the financial constraint in which many policy makers find themselves operating.

Finally, one of the limits of models often highlighted by policy makers is their relative complexity – be a model intended as a mathematical representation of reality, or as a process of building a mental map of reality. It is therefore quite tempting for researchers to simplify their tools and methodologies, with however the risk that excessive simplification will lead to biased – or outright wrong – results and advice. Simplification of existing tools and methodology should therefore be seen only as a second best solution, while layered modelling systems, meta-models and improved interfaces and communication tools should be preferred.

7.5 Conclusions

Integrated Water Resource Management requires public participation as one fundamental component of sustainable management of water resources, enabling the integration of the social and political spheres of resource planning. The requirement for participation in planning and management is also enshrined in the main European references and initiatives, such as the WFD and the MED EUWI.

The implementation of the declared principles in the practice of IWRM plans however imposes on policy makers the burden of undertaking consultation and participatory planning and management, without providing adequate operational methods and tools for implementing it. This is particularly true in the Mediterranean area, where the practice of PP is relatively newer than in northern European countries. Hence, policy makers are left with a three-faceted problem: first of all, how to select all relevant stakeholders in an objective and transparent way and to do so in a manner which ensures results are a true reflection of the public, as well as respecting sustainability criteria. Secondly, how to constructively involve the identified stakeholders – hence managing their inter-relations, interests, etc. And, finally, how to integrate their opinions, concerns, desires in the planning process together with technical and scientific information coming from various sources, such as monitoring systems and, in particular, various kinds of disciplinary simulation models, in a manner which is transparent and manages to compromise among objectives and to build consensus.

Models are often cited as one possible avenue to improve water resources management and planning. Within the broader meaning of the term, models include mental maps and external representations of internal interpretations of reality – thus moving away from hard-fact, objective formalisations to include the process of developing models and building tools. Participatory modelling can provide a means to link mental models to mathematical models, and DSS's are the tools which can be used to explore the iterative process within the paradigm of participated decision making. The process of developing DSS models – understood to encompass a broad range of aspects, that go beyond mere physical models mimicking ecological or operational processes to include the integration of social and economic dimensions – has proven useful in shaping concerted policies, facilitating actors' involvement in the process, and reaching compromises on conflicting issues. Further strengthening this view is the realisation that traditional DSS focusing on modelling and the computation part of decision process have consistently failed to consider the "soft part" of the decision-making process, and thus problem structuring and conflict mitigation. These latter activities cannot be translated into computer codes, but are of course integral and important parts of problem management and decision making, and can be integrated within a participatory modelling framework.

It is thus the process of participatory modelling which can more readily and effectively support participatory planning for IWRM and, in the European context, for the WFD. Yet until research and policy find a common ground, this potential will remain untapped. Some general lessons can be drawn from the experience described in this paper:

- There is a strong need for scientific and technical support for the meaningful involvement of stakeholders in planning and implementation through participatory modelling
- Models which can facilitate public participation need to be understood in a much broader way, as components of the process of engaging both decision makers and the general public to make participation worthwhile
- To bridge the gap between models as tools and actual decisional processes, one need to move on to the next stage of "model" development, where the tool is developed with heavy involvement of end-users and stakeholders through participatory modelling, and it is geared to addressing their priority needs; yet, there is a trade-off between specificity and re-usability of both participatory planning methodologies and the resulting tool, thus requiring flexible and, possibly, modular approaches to be developed.

The experience of MULINO represented one of the first attempts to address some of the issues mentioned, demonstrating however that the barrier between research and policy has not been broken yet. The challenge remains of how to trigger the process of collaborative development and adoption of participatory modelling methodologies and resulting DSS tools by potential end users. Mutual learning between scientists and policy makers may then help to make the process of participatory modelling general enough to be applicable in a diversity of contexts, and with different models, but at the same time with enough scientific robustness to ensure its usefulness in structuring the problem and identifying a range of possible solutions. Such a methodology, with the support of integrated modular tools, should cover all aspects of participatory management – from the identification of the relevant stakeholders to the methods for integrating their concerns in the final choice – and would provide sound advice and technical/scientific support for policy makers, to implement a concept as elusive as that of participatory planning, which remains a requirement of the IWRM. Otherwise, the risk is that participation will remain just another "buzz-word", with the consequent failure of harnessing its huge potential benefits in terms of improved decision making.

Acknowledgments Research carried out with partial financial support of the European Community projects MULINO (EVK1-2000-22089), TRANSCAT (EVK1-CT-2002-00124), NOSTRUM-DSS (INCO-CT-2004-509158), and of the Italian Ministry for University and Research (PRIN ex 40% 2004 and FIRST ex 60% 2005) under the scientific direction of Prof. Carlo Giupponi.

References

Ackoff, R.L. (1979). The future of operational research is past. *Journal of Operational Research Society*, 30: 93–104.

Allen, W., Kilvington, M., and Horn, C. (2002). Using participatory and learning-based approaches for environmental management to help achieve constructive behavioral change. *Land-Care Research Contract Report* 102.

Arnstein, S.R. (1969). A ladder of citizen participation, *Journal of American I Planners*, 35: 216–224.

Cooke, B. and Kothar, U. (2001). *Participation: the New Tyranny?* Zed Books, London.

Dalal-Clayton, D.B. and Bass, S. (2002). *Sustainable Development Strategies: A Reference Book*. Earthscan Publications, London.

Doyle, J.K. and Ford, D.N. (1998). Mental models concepts for system dynamics research, *System Dynamic Review*, 14: 3–29.

Dyson, R. and O'Brien, F. (1998). *Strategic Development: Methods and Models*. Wiley, Chichester, UK.

EC European Commission (2003a). EC Common Strategy for the Water Framework Directive (2000/60/EC). Guidance Document No. 8: Public participation in relation to the Water Framework Directive. Active Involvement, Consultation and Public Access Information. Office for the Official Publications of the European Communities, Luxembourg.

EC European Commission (2003b). *EC Common Strategy for the Water Framework Directive (2000/60/EC). Guidance Document No. 11: Planning Process*. Office for the Official Publications of the European Communities, Luxembourg.

EEA European Environmental Agency (1999). *Environmental Indicators: Typology and Overview*. European Environmental Agency, Copenhagen, 25 pp.

Feás, J., Giupponi, C., and Rosato, P. (2004). Water Management, Public Participation and Decision Support Systems: the MULINO Approach, Complexity and Integrated Resources Management. *Transactions of the 2nd Biennial Meeting of the International Environmental Modelling and Software Society* 2: 617–622.

Flood, R. and Jackson, M. (1991). *Creative Problem Solving*. Wiley, London.

Funtowicz, S.O. and Ravetz, J.R. (1993). Science for the post-normal age, *Futures*, 25: 739–755.

Geurts, J.L.A. and Joldersma, C. (2001). Methodology for participatory policy analysis, *European Journal of Operational Research*, 128: 300–310.

Giupponi, C. (2006). Decision support systems for implementing the European water framework directive: the MULINO approach. *Environmental Modelling and Software*, in press.

Giupponi, C., Camera, R., Lasut, A., Mysiak, J., and Sgobbi, A. (2005). Facilitating participatory decision processes in integrated river basin management. In: *6th International Conference of the European Society for Ecological Economics, June 14–17, 2005*, Lisbon. EESE. Oral communication.

Giupponi, C., Mysiak, J., Fassio, A., and Cogan, V. (2004). MULINO-DSS: A computer tool for sustainable use of water resources at the catchment scale. *Mathematics and Computers in Simulation*, 64: 13–24.

Giupponi, C., Jakeman, A.J., Karssenberg, D., and Hare, M.P. (2006). *Sustainable Management of Water Resources: An Integrated Approach*. Edward Elgar, Cheltenham, UK.

GWP-TAC (2000). *Integrated Water Resources Management*. Global Water Partnership (GWP), Stockholm, TEC Background Paper No. 4

Hailey, J. (2001). *Beyond the Formulaic: Process and Practice in South Asian NGOs*. Zed Books, London, 88–101.

Hare, M. (2004a). Policy Makers' requirements for models to support the participatory element of the water framework directive. Harmoni-CA Project, Brussels.

Hare, M. (2004b). Modellers' recommended research directions for models to support the participatory elements of the water framework directive. Harmoni-CA Project, Brussels.

Hare, M. (2005). The use of models to support the participatory elements of the EU water framework directive: Creating a dialogue between Policy Makers and Model Makers. Harmoni-CA Project, Brussels.

Hare, M., Letcher, R.A., and Jakeman, A.J. (2003). Participatory modelling in natural resource management: A comparison of four case studies, *Integral Assessment Journal*, 4: 62–72.

Hare, M.P., Barreteau, O., Beck, M.B., Letcher, R., Mostert, E., Tabara, J.D., Ridder, D., Cogan, V., and Pahl-Wostl, C. (2006). *Methods for Stakeholders' Participation in Water Management*. Edward Elgar, Cheltenham, UK.

Hildyard, N., Phedge, P., and Wolvekamp, P. (2001). *Pluralism, Participation and Power: Joint Forest Management in India*. Zed Books, London, 56–71.

Hodgson, A.M. (1992). Hexagons for systems thinking. *European Journal of Operational Research*, 59: 220–230.

ICWE, (1992). The Dublin Statement, International Conference on Water and Environment.

Letcher, R. and Bromley, J. (2005). *Typology of Models and Methods of Integration*. Edward Elgar, Cheltenhan, UK, 287–323.

Lourenço, N., Rodrigues L., and Russo Machado, C. (2004). Social networks and water management decision making: a methodological approach to local case studies, *MULINO Project Deliverable*.

McIntosh, B.S., Jeffrey, P., Lemon, M., and Winder, N. (2004). On the design of computer-based models for integrated environmental science, *Environmental Management*, 35: 741–752.

Mendoza, G.A. and Prabhu, R. (2005). Combining participatory modelling and multi-criteria analysis for community-based forest management, *Forest Ecology and Management*, 107: 145–156.

Mingers, J. and Rosenhead, J. (2004). Problem structuring methods in action. *European Journal of Operational Research*, 152: 530.

Mosse, D. (1994). Authority, gender and knowledge: theoretical reflections on the practice of participatory rural appraisal, *Development and Change*, 25: 197–526.

Mysiak, J. (2005). Decision support systems for integrated water resource management: value and success factors. *Proceedings of International Symposium on Environmental Software Systems*.

Mysiak, J., Giupponi, C., and Rosato, P. (2005). Towards the development of a decision support system for water resource management, *Environmental Modelling Software*, 20: 203–214.

Nelson, N. and Wright, S. (1995). *Power and Participatory Development: Theory and Practice*. IT Publications, London.

Pimbert, M. (2004). Institutionalising participation and people-centered processes in natural resource management. International Institute for Environment and Development, London.

Renn, O. (2006). Participatory processes for designing environmental policies. *Land Use Policy*, 23: 34–43

Richards, M., Davies, J., and Cavendish, W. (1995). Can PRA methods be used to collect economic data? A non-timber forest product case study from Zimbabwe. *PLA Notes*, 36: 34–40.

Rittel, H. W. J. and Webber, M. M. Dilemmas in a general theory of planning, *Policy Science*, 155–169.

Rosenhead, J. (1989). *Rational Analysis for a Problematic World*. Wiley, Chichester, UK.

Rosenhead, J. and Mingers, J. (2001). *Rational Analysis for a Problematic World Revisited*. Wiley, Chichester, UK.

Sgobbi, A. and Fraviga, G. (2006). Governance and Water Management: Progress and Tools in Mediterranean Countries. FEEM Working Paper No. 151.06 Available at SSRN: http://ssrn.com/abstract = 951461.

Stirling, A. (2006). Analysis, participation and power; justification and closure in participatory multi-criteria analysis. *Land Use Policy*, 23: 95–107.

Swyngedouw, E., Page, B., and Kaika, M. (2002). Achieving participatory governance: sustainability and policy innovation in a multi-level context. *Cross-Cutting Issues in the Water Sector*. Oxford.

Taylor, H. (2001). Insights into Participation from Critical Management and Labour Process Perspectives. *Zed Books*, London, pp. 122–138.

UN (1992). United Nations Convention on the protection and use of transboundary water courses and international lakes (Helsinki Convention).

UN (1993). Agenda 21: Earth Summit – the United Nations Programme of Action from Rio. United Nations, New York.

UN (1998). Convention on Access to Information, Public Participation in Decision-Making, and Access to Justice in Environmental Matters (Aarhus Convention).

van der Sluijs, J.P. (2002). A way out of the credibility crisis of models used in integrated environmental assessment, *Futures*, 34: 133–146.

Vennix, J.A.M. (1996). *Group Model Building: Facilitating Team Learning Using System Dynamics*. Wiley, Chichester, UK.

Vennix, J.A.M., Gubbels, J.W., Post, D., and Poppen, H.J. (1990). A structured approach to knowledge elicitation in conceptual model building. *System Dynamics Review*, 31–45.

Walters, C. (2002). *Adaptive Resource Management*, The Blackburn Press.

Wasserman, S. and Faust, K. (1994). *Social Network Analysis: Methods and Applications*. Cambridge University Press, New York.

Zapatero, E.G. (1996). A quality assessment instrument for multi-criteria decision support software, 17–27(11).

Part II
Results from the Developing World

Chapter 8
Evaluating the Institution-Impact Interactions in the Context of Millennium Development Goals

Analytical Framework with Empirical Results*

R. Maria Saleth, Ariel Dinar, and Susanne Neubert**

8.1 Introduction

Governments and development agencies constantly plan, implement, and evaluate various development interventions, i.e., projects, programs, and policies. These interventions vary in scale and coverage, ranging from those specific to a group, region, resource, or sector to those universal and global in scope. Considering the flow and magnitude of investments involved, there is an understandable concern over the actual impacts that these interventions generate. Despite this concern, two key aspects having a central role in determining the magnitude and sustainability of development impacts continue to lack recognition and treatment both in economic literature and in development policy. These are: (a) the role institutions play in impact generation and transmission and (b) the synergies inherent among past, ongoing, and planned interventions. The insufficient treatment of institutional roles and the failure to account for development synergies could create fundamental bias in development planning and impact assessment. This problem is particularly serious in the context of meta-development goals such as the Millennium Development Goals (MDGs), where the realization of the final goal is linked with the realization of several intermediate, but related goals of a hierarchy of development interventions,

* Presented at the International Seminar Water Management: Technology, Economics and the Environment, Organized by the Foundation Ramon Areces, Madrid, January 19–20, 2007. The empirical component of this chapter was supported by Seenithamby Manoharan, Sarath Abayawardana, Ranjith Ariyaratne, and Bandi Kamaiah. This chapter is a product of a study *The Institutional Matrix of the Millennium Development Goals: An Empirical Study of Food Security Goals in Kala Oya Basin, Sri Lanka* funded by the World Bank Research Committee and International Water Management Institute. The views expressed here are those of the authors and should not be attributed to the World Bank.

** The authors are affiliated respectively with the International Water Management Institute, Colombo, Sri Lanka, World Bank, Washington DC, USA, and German Development Institute, Bonn, Germany.

P. Koundouri (ed.), *Coping with Water Deficiency,* 189–212.

all of which require an effective institutional framework for their implementation and monitoring.

Since the MDGs are now treated as an accepted framework for development planning and tools for monitoring development progress [United Nations (UN), 2006], the issue of capturing the facilitative effects of institutional roles and impact synergies assume practical importance and current relevance. Unfortunately, this issue is neither a part of the framework nor a part of the tools that the MDG administration (UNDP, 2006) has developed to support the design, evaluation, and monitoring of the MDG-oriented development interventions. This gap also persists in the small but growing literature aiming to assess the progress on MDGs (e.g., World Bank and IFPRI, undated; Sahn and Stife, 2002; Haines and Cassels, 2004). Since these studies only extrapolate the future progress based on *ex-post* performance up to a given year, they fail to incorporate the *ex-ante* dimension of what would happen when institutional performance is enhanced and development synergies from completed, ongoing, and planned interventions are reckoned.[1] This paper aims to fill this gap by developing a methodology that could directly capture both institutional impacts and development synergies within a unified framework and quantitative context. The methodology is demonstrated by taking food security related to MDG1 as an example, the Kala Oya Basin in Sri Lanka as the empirical setting, and stakeholder-based *ex-ante* qualitative information as the data source.

From here on, the paper is structured as follows. Section 8.2 discusses the welfare impact from policy intervention and the value of its *ex-ante* evaluation. Section 8.3 sets the conceptual foundation and analytical framework of the proposed methodology and describes the institution-impact matrix. Section 8.4 describes the empirical context and data generation. Section 8.5 applies the institution-impact matrix to the development and institutional context of the study region. Section 8.6 presents and analyzes the results of the econometric models of institution-impact interaction and illustrates the role of institutional impacts and development synergy. The final section concludes with analytical and empirical insights as well as limitations of the approach and scope for further extension and refinement.

8.2 *Ex-Ante* Policy Assessment-A Welfare Perspective

When selecting policies, policymakers usually consider their effects not only on the total welfare but also on its distribution across groups.[2] But, the *ex-ante* issue of what would happen to this welfare and its distribution when the roles of relevant

---

[1] For example, Haines and Cassels (2004) have first estimated a trend line with ex-post data for two years and, then, used this to extrapolate the future trend till the target date of 2015. The distance between these two trend lines is taken as the gap between the actual and desired levels of achievement. This approach is also followed by World Bank and IFPRI (no date provided).

[2] The distributional impacts are particularly important in policies, such as the MDGs, which, by their nature, target the special and laggard population groups.

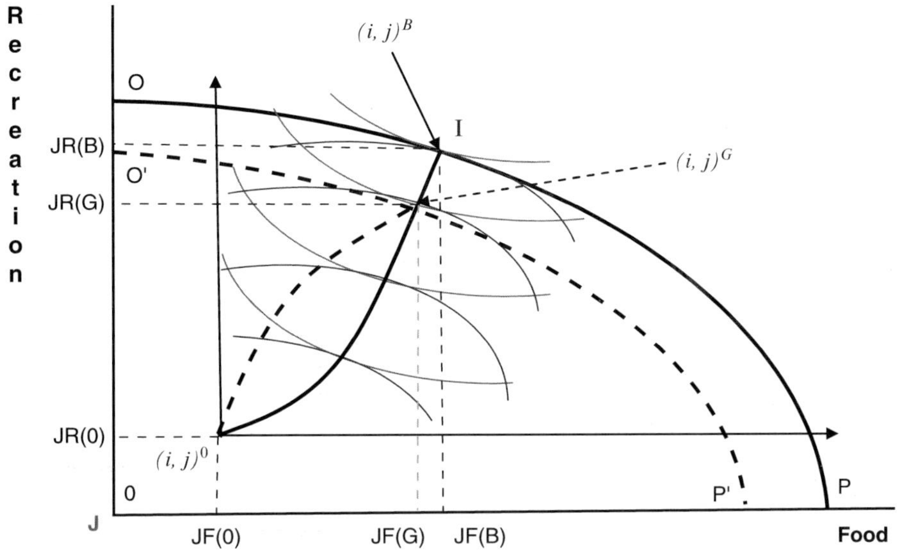

Fig. 8.1 Evaluation of alternative policy paths and societal welfare

institutions and synergies of related policies is ignored. The policy value of the *ex-ante* consideration can be graphically demonstrated using Fig. 8.1, which is an adaptation of a framework suggested by Just et al. (2004).

Figure 8.1 depicts a simple economy with two individuals (or groups), I (rich) and J (poor), who, with a given bundle of resources, can produce/consume two goods, i.e., food (F) and recreation (R). Given current technologies and institutions in the economy, the production possibility frontier of the economy is OP. Assume further that the economy is in a status quo at $(i, j)^0$ with a corresponding welfare level for the two-person society. J's welfare is JF(0) + JR(0) and I's welfare is [P-JF(0)] + [O-JR(0)]. Assume that this welfare distribution is not equitable and the government considers intervention, aimed at improving the welfare of J. Of course, the government policy aims to remain on OP. The government considers two policies that could *a priori* achieve such economic and social objectives, i.e., a 'dashed' (dashed line) policy intervention and a 'solid' (solid line) policy intervention, which are expected to change welfare allocation to improve J's (the poor) situation.

Assume that the policymaker knows the path and the destination of the economy, resulting from each policy intervention. The 'dashed' policy intervention moves the economy from $(i, j)^0$ to $(i, j)^G$, and the 'solid' policy intervention moves the economy to $(i, j)^B$. Both policy interventions are Pareto optimal (satisfying utility maximization of the two individuals), but the 'dashed' policy is less efficient as it falls short of the production possibilities frontier (OP) and ends with an inner frontier, O'P' < OP. While the 'dashed' policy intervention is less efficient, it may be politically less controversial, as pressure from individual I (rich) may be less due to his bigger share of allocated society resources compared to that under the 'solid' policy intervention.

Such analysis is very useful if the policymaker also knows and accounts for the possible technical and economic externalities from institutional facilitations and impact synergies. The policymaker gains substantial political and practical credits from this *ex-ante* information.

It can be shown that when the policymaker now considers the institutional roles and impact synergies, both higher welfare and its preferred distribution can be achieved. For example, if an irrigation-based development program is implemented in conjunction with the introduction of an institution aiming to improve water distribution, not only the production possibility frontier but also the development path will be different. The resultant equilibrium can enhance both the level and distribution of welfare. In a similar vein, it is also possible to show the welfare gain from incorporating impact synergy among development interventions. For example, when the irrigation development is accompanied, say, by a program of agricultural intensification, the higher levels of output and employment will not only enhance total welfare but also improve its distribution. While it is easier to conceptualize the welfare consequences of institutional roles and impact synergies, it is a major methodological challenge to analytically capture and empirically assess these effects. In the following sections, we demonstrate one approach that addresses this challenge.

8.3 The Analytical Framework

The work of Saleth and Dinar (2004) is extended to develop an analytical framework needed for explicitly accounting for the role of institutional impacts and development synergies.[3] The building blocks of this framework are: the *institutional ecology* principle, the *institutional decomposition and analysis* (IDA) approach [similar to that of E. Ostrom (1990)], the *ex-ante* approach, and the *adaptive instrumental evaluation* (Tool, 1977; Kahneman and Tversky, 1984; Bromley, 1985). While these concepts are explained in Saleth and Dinar (2004), here we indicate how they are used to set the analytical framework for evaluating the institution-impact interaction.

The institutional ecology principle enables one to view basin institutions as a nested and interlinked system embedded within a given physical, social, and political economy context. The IDA framework allows an analytical unbundling of basin institutions (i.e., water, land, agricultural, and environmental) to identify their key components, show the structural and functional linkages among them, and trace the relevant institutional configurations operating beneath various impact pathways of different development interventions. As we will show later, the adaptive instrumental evaluation is used to get perception-based ex-ante qualitative information on institutions and impacts from stakeholders. The development of the analytical framework

[3] A general application of this framework for a global ranking of institutional health and reform prospects within the water sector is illustrated in Dinar and Saleth (2005).

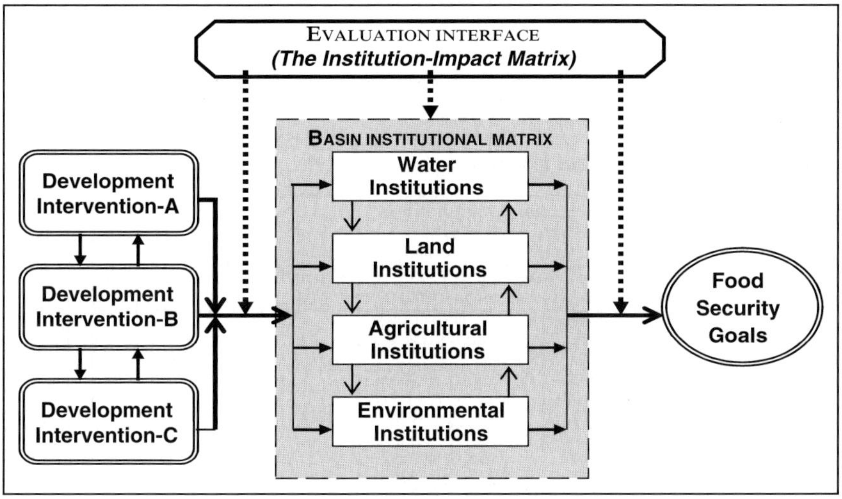

Fig. 8.2 Conceptual frame for institution-impact interface

begins first with the simple conceptualization of the relationships among the development interventions, institutional configurations, and food security goals. The basic conception of the model of institution-impact interaction is shown in Fig. 8.2.

To operationalize this conceptual model, the original methodology of Saleth and Dinar (2004), which was developed for the particular context of institution-performance interaction within water sector, requires some adjustments. First, institutional evaluation is to be specialized to a local region (e.g., basin), where it is possible in order (a) to identify completed, ongoing, and planned development interventions, (b) to trace their major impact pathways, and (c) to map all the relevant institutions involved in various points of these paths. Second, the evaluation is to be extended to cover not just water institutions but also the land, agricultural, and environmental institutions, including those needed to cope with the water and food consequences of various policy and external shocks, such as climatic change, within an integrated framework. The focus is as much on the individual performance of these institutions as on their collective performance as reflected in terms of their structural and operational linkages (North, 1990; Saleth and Dinar, 2004). And, third, the evaluation has also to be performed within the framework of a multidimensional institution-impact matrix. The derivation of this matrix is illustrated below.

8.3.1 Institution-Impact Matrix

The institution-impact matrix captures the operational relationships and synergy among development interventions, impact pathways, institutional configurations, and food security goals. For illustrating the derivation of the institution-impact

Development Intervention 3 (Watershed Development)

Development Intervention 2 (Introduction of New Crop Varieties)

Development Intervention 1 (Water Development Projects-Dams)

Impact Pathways	Basin Institutional Configurations (Defined by combinations of variables capturing the land, water, agricultural, and resource/environment institutions)				Development Goals (In terms of variables capturing poverty, food and resource conservation goals)		
	Water Institutions	Land Institutions	Agricultural Institutions	Res/Env Institutions	Income/ Jobs	Food Price/ Output	Efficient Use of Resources
Irrigation	Water rights/ User organs ($$$)	Land tenure/ Tenancy ($$$)	Input system/ extension ($$$)	Water/soil quality codes ($$$)	($$$)	($$$)	($$$)
Regional Growth	Basin institutions ($$$)	Land markets ($$$)	Farm wage policy ($$$)	Pollution regulations ($$$)	($$$)	($$$)	($$$)
Urbanization	Sectoral water Allocation ($$$)	Farmland use rules ($$$)	Rural-urban markets ($$$)	Watewater regulations ($$$)	($$$)	($$$)	($$$)
Water Supply	Pricing & cost recovery ($$$)	Property ownership ($$$)	Wastewater use practice ($$$)	Sanitation Policy ($$$)	($$$)	($$$)	($$$)
Ecological Effects	Project choice & scale ($$$)	Policy on commons ($$$)	Policies on fragile areas ($$$)	Forest laws & policies ($$$)	($$$)	($$$)	($$$)

Note: $$$ = Values of actual or perception-based information on one or more variables designed to capture both the status and effectiveness of each of the institutional aspects as well as the level of impacts on each of the food security aspects.

Fig. 8.3 Institution-impact matrix, a simplified presentation

matrix, we take three development interventions, i.e., water development, crop variety introduction, and land/soil improvement, and the development goal of food security. With the identification of the impact pathways and possible institutional configurations involved in these pathways and with the delineation of the income, price, and resource components of the food security goal, the conceptual framework in Fig. 8.2 can be operationalized in terms of the institution-impact matrix (Fig. 8.3).

While the institutions specified in the case of different impact pathways are not exhaustive but only illustrative, it shows clearly how different institutional configurations are involved in impact transmission. It can also be noted that even though the institutional configuration involved in a given impact pathway is the same, the relative role of individual institutions can be different depending on the three sub-components of the development goal. As a result, each row of the matrix, in fact, represents three separate but related institution-impact relations. These relations for all the rows corresponding to each of the three development programs can, therefore, be translated into an empirically testable set of relationships (equations), allowing capturing the underlying institutional and impact aspects. In Section 8.4 we identify an empirical context where the methodology is illustrated with real-life conditions of development. But, before that it is instructive to see the stylized set of institution-impact interaction implicit in Fig. 8.3.

8.3.2 A Stylized System of Interactions

The various components of the Institution-impact can be expressed in a set of equations that reflect the interactions between the policy variables, the existing institutions, the interim impacts, and the ultimate development goal. Once the equations of such a system are estimated, the coefficients of the various variables provide important information to policymakers. One piece of information derived from the empirical results is the relative importance of policy interventions and institutions in the development process, and thus, the synergy among them. This will be discussed in the results section. Another piece of information that is embedded in the sequential/simultaneous nature of the interactions can provide information on the relative effectiveness of channels of impact of policy interventions that could quite similarly be traced to the discussion around Fig. 8.1.

Here, we provide a stylized discussion on how one can transform the institution-impact system coefficients into a guide for policymakers. For showing this, let the said system be represented by four sets of equations:

$$G = g(\underline{M}_g) \tag{1}$$

$$\underline{D}_d = d(\underline{D}_d^{-p}, \underline{N}_d) \tag{2}$$

$$\underline{M}_m = m(\underline{N}_m, \underline{D}_m, \underline{M}_m^{-p}) \tag{3}$$

$$\underline{N}_n = n(\underline{N}_n^{-p}, \underline{D}_n) \tag{4}$$

Let $G = g(\underline{M}_g)$ be the single equation representing the link between impact variables $(\underline{M}_g)$ and the MDG Goal (G). Let $\underline{D}_d = d(\underline{D}_d^{-p}, \underline{N}_d)$ be the equations representing the link between a set of Development interventions $\underline{D}_d^{-p}$ and a set of institutions $(\underline{N}_d)$ on the Development interventions in the vector of dependent variables $(\underline{D}_d)$. Let $\underline{M}_m = m(\underline{N}_m, \underline{D}_m, \underline{M}_m^{-p})$ be the linkage between a set of Development interventions $\underline{D}_m$, a set of institutions $(\underline{N}_m)$ and a set of some impact variables $(\underline{M}_m^{-p})$ on the development interventions in the vector of dependent variables $(\underline{M}_m)$. And finally, let $\underline{N}_n = n(\underline{N}_n^{-p}, \underline{D}_n)$ be the linkage between a set of Development interventions $\underline{D}_n$, a set of iNstitutions $(\underline{N}_n^{-p})$ on the set of institutions in the vector of dependent variables $(\underline{N}_n)$. The variable $\underline{X}_x^{-p}$ $\forall X = D, M, N;\ x = d, m, n$ is the vector of variables in $\underline{X}_x$ not including the dependent variable of that equation. The number of equations in this system is $1 + d + m + n$. Because of the sequential and simultaneous nature of the system, it is easy to show that it can be reduced into one equation $G = g(m(n(\underline{N}_n^{-p}, \underline{D}_n), d(\underline{D}_d^{-p}, n(\underline{N}_n^{-p}, \underline{D}_n)), \underline{M}_m^{-p}))$ that can allow the analyst deriving, using the chain rule, the relationship between the development goal and each of the sets of variables. We will not perform the analysis in this paper, but it is sufficient to say that the ultimate results will allow a follow-up on every variable and its relationship to any other variable within the system.

8.4 The Empirical Context: The Kala Oya Basin, Sri Lanka

We apply the institution-impact assessment framework to the institutional and development context of the Kala Oya Basin in Sri Lanka (Fig. 8.4). The Kala Oya Basin, which is one of the 108 basins in Sri Lanka, covers an area of 2,873 square kilometers and supports a population of about 0.41 million. Of the total land area of 287,303 hectares (ha), far less than a third is cultivable due to land slope and water-related constraints. Paddy cultivation and home gardens with coconuts and fruit trees account for 40 percent of the cultivated area (de Silva et al., 2006). Water scarcity is also serious due to the seasonal patterns of rainfall and water quality problems. Increasing population density and aging are the main the demographic issues. The incidence of poverty remains substantial in the basin. For example, in the Anuradhapura district, which accounts for half of the basin area, the percentage of people below the official poverty line (Rs. 1,423 or US$14/capita/month) is reckoned at 20% during 2000–01 (de Silva et al., 2006), and 44% of the families in the basin rely regularly on *Samurdhi*, the poverty reduction program of the government. Food insecurity is also serious, as many villages in the basin area fall under the most vulnerable categories of food insecurity (DCS and WFP, 2005). A more detailed review of the Basin's poverty level and the strategic reasons for its selection for our case can be found in Saleth et al. (2006).

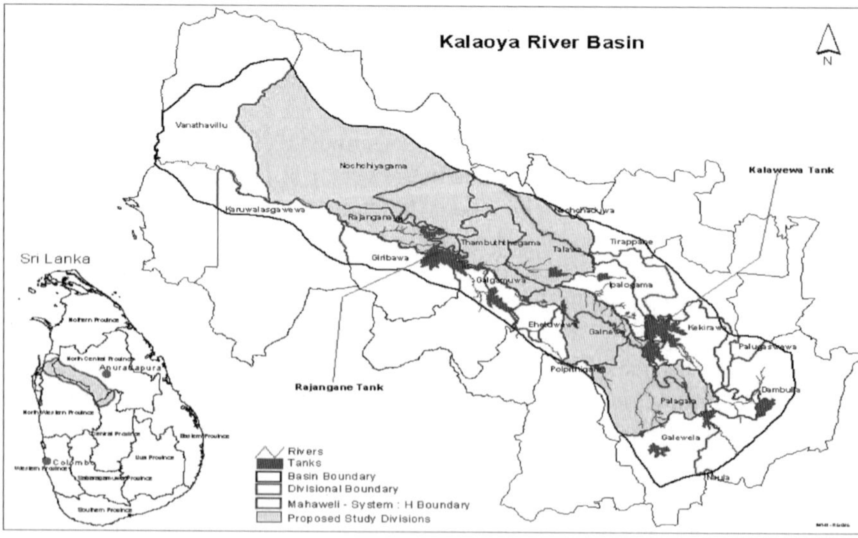

Fig. 8.4 The Kala Oya Basin, Sri Lanka

8.5 Empirical Specification of the Model

For the empirical translation of the matrix in Fig. 8.3 (or equations [A]–[D]), we need to identify the development goal, development interventions, and the relevant set of institutions. Considering the conditions of the study basin, we take food security, a key component of MDG1, as the development goal. As to the development interventions, we consider three, namely, crop diversification, system rehabilitation, and bulk water distribution.[4] It is now possible to trace and delineate the major pathways through which these interventions may impact on food security. Given these impact pathways, it is also possible to identify the set of institutions (i.e., agriculture, water, and land-related legal, policy, and organizational aspects) that are likely to affect the generation and transmission of impacts with and across pathways. Figure 8.5 depicts these impact pathways and their underlying institutional configurations associated with the three development interventions.

Figure 8.5 is very instructive as to the development synergies among the interventions as well as the specific point at which different institutions influence the impact flows. Although Fig. 8.5 needs to be read from left to right in line with the direction of transmission chains and impact flows, for analytical convenience, it is useful to move recursively, i.e., starting with the immediate variables affecting food security, then, tracing the variables affecting these intermediary variables and so on. In doing so, we could identify many impact pathways and their associated chains of development, impact, and institutional aspects. If we define a set of variables to capture these aspects, then all the institution-impact interactions in Fig. 8.5 can be mathematically represented as a system of linked equations (see Section 8.3.2). To show this, we define 32 variables listed in Table 8.1.

As can be seen from Table 8.1, the variables cover one development goal variable and three intervention, 17 impact, and 11 institutional variables.[5] Obviously, the variables differ considerably in terms of their unit of measurement, evaluation domain, amenability for observation, and scope for getting actual data. To avoid the problems due to their diverse features, we conceive all the variables essentially in a qualitative sense to be evaluated on an interval of 1–10, with 1 being the lowest

[4] Of these three, system rehabilitation has already been implemented, while bulk water distribution is being implemented only on a pilot scale in the canal irrigation system of the basin. Crop diversification, is only being planned, though the Government of Sri Lanka has a national policy for promoting high-value crops.

[5] These are actually the economic and technical factors that act as the 'impact transmission variables'. They are denoted here as 'impact variables'. They should not be confused with those in the impact assessment literature, where 'impact variables' refer only to the ultimate end-goals. 'Impact variables' in this paper are treated as 'outcome variables' in that literature (Neubert, 2000). In the context of our framework, it is still appropriate to treat them as impact variables because (a) they do capture the intermediary impacts (or, outcomes) and (b) such impacts are specifically evaluated using equations representing different impact layers.

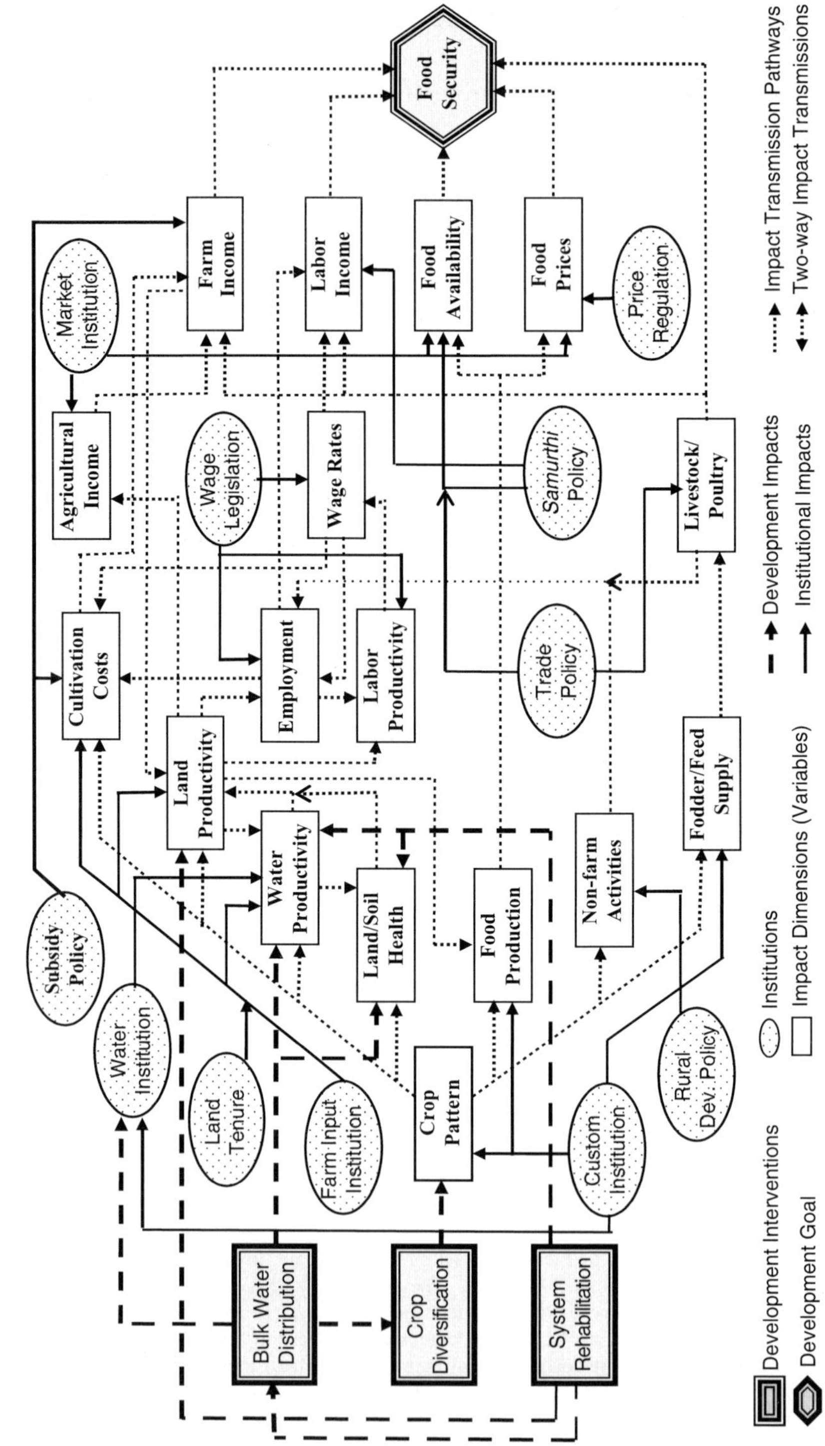

Fig. 8.5 Institution-impact system for food security with three development interventions

Table 8.1 Variables in the institution-impact-impact model

Symbol	Categories of variables	No	Names of variables	Acronym used
G	Development goal	1	Food Security	FODSECTY
D	**Development interventions**	**1**	**Crop Diversification**	**CROPDIVR**
		2	System Rehabilitation	SYSREHAB
		3	Bulk Water Distribution	BULKWATD
M	Impact variables	1	Crop Pattern	CROPATEN
		2	Land Productivity	LANPRODY
		3	Water Productivity	WATPRODY
		4	Labor Productivity	LABPRODY
		5	Rural Employment	RURALEMP
		6	Wage Rates	WAGERATE
		7	Cultivation Costs	CULTCOST
		8	Agricultural Income	AGLINCOM
		9	Land Quality/soil Health	LANHELTH
		10	Food Production	FODPRODN
		11	Non-farm Enterprises	NFAMENTS
		12	Fodder & Feed Supply	FEDSUPLY
		13	Livestock/Poultry	LIVSTOCK
		14	Farm Income	FAMINCOM
		15	Wage Income	LABINCOM
		16	Food Availability	FODAVAIL
		17	Food Price	FODPRICE
N	Institutional Variables	1	Land Tenure	LANTENUR
		2	Water Institutions	WATINSTN
		3	Farm Input Institutions	FAMINSTN
		4	Customary Institutions	CUSINSTN
		5	Rural Development Policy	RDVPOLCY
		6	Market Institutions	MKTINSTN
		7	Wage/Labor Legislations	WAGELAWS
		8	Trade Policy	TRDPOLCY
		9	Price Regulations	PRICREGL
		10	Farm Subsidy Policy	SUBPOLCY
		11	Samurthi Policy	SAMPOLCY

and 10 being the highest.[6] In this format, the variables capture only the overall perception as to their status, change, effectiveness, or impact. For example, the food security variable represents only an overall perception as to its overall status taking implicitly, the adequacy and quality of food consumption across groups.[7] Similarly, the variables representing the interventions are considered to capture their overall effectiveness or impact potential.

[6] Such an approach also enables us to circumvent the non-availability of getting observed data by tapping the knowledge of stakeholders with a carefully designed survey instrument.

[7] It is considered to be affected by four proximate variables, i.e., income, food prices, food availability, and self-consumption possibilities, particularly of home grown livestock/poultry products.

Institutional variables capture the status, effectiveness, or impact of institutions with respect to different impact pathways and contexts. For example, the variable LANTENUR captures the conduciveness of land tenure (farm size and ownership) to crop pattern changes, land productivity, etc. The impact variables capture the actual or expected changes due to the impacts of interventions and institutions in different contexts of impact generation and transmission. Among the income variables, distinction is made between farm income (covering agricultural and livestock incomes) and labor income (covering wage and livestock income) to capture the differential income potentials between those with and without access to land. Given the set of variables listed in Table 8.1, the institution–impact framework in Fig. 8.5 can be formally represented in a mathematical form with a set of 21 equations that comprise the system model of institution–impact interaction.

It can be verified that each of these equations correspond to one of the 21 impact pathways evident in Fig. 8.5. As we look into the equations, it is clear that they are structurally linked in view of the sequential (in most cases) as well as simultaneous (in few cases) relationships among them. The equations are arranged sequentially, starting with the initiation of the development interventions, then, with their impacts in the order of their occurrences, and finally, ending with the impact on the ultimate development goal, i.e., food security. Thus, the order in which the equations are sequenced captures the relative position of different layers within the upstream-downstream continuum of impact transmission. At the same time, the configuration of variables in each equation is based on two considerations: (a) the functional relationship expected between them and the independent variable as per economic reasoning and (b) the need for avoiding linkages among independent variables to minimize the scope for the econometric problem of multicollinearity.[8] Given the functional linkages among variables and sequential linkages among equations, the impact and institutional variables can be hierarchically arranged by tracing their role and positions both within and across the impact pathways.

Of the 32 variables in the analysis, the 11 underlined variables are independent or exogenous (includes all the institutional variables except one, i.e., water institution—WTINSTN, and also one of the development interventions—SYSREHAB). But, the remaining 21 variables are dependent or endogenous covering 17 impact variables, two development variables representing respectively the two interventions of CROPDIVR and BULKWATD, and one institutional variable representing WATINSTN. Given the way all the 21 equations are specified in terms of the configuration of endogenous and exogenous variables, they satisfy both the rank and order conditions necessary for their identification and unbiased estimation (Johnston, 1984; Kennedy, 1987).

---

[8] These two considerations can be at odd because the economic consideration can warrant the inclusion of one or more independent variables, even though they may be closely related. Whether this leads to the econometric problem of multicollinearity can be tested using (a) correlation analysis of the independent equations and (b) indicators such as very high R^2, low t-ratio, and changing signs of some key variables. Note that the multicollinearity problem will neither affect the model fit nor the efficiency properties of the estimated coefficients.

8.6 Data and Preliminary Results

While the structural model defined by the system of 21 equations is econometrically consistent and intuitively appealing, it has a major empirical challenge because consistent and comparable data on both the impact and institutional variables are difficult to get. It is certainly possible to get actual data on some of the impact variables (e.g., productivity, employment, income, and wage rates) through, for example, a household survey. However, the information thus collected will represent only the past impact of an already implemented development intervention and could not capture the synergy from the expected impacts of ongoing and planned intervention. Still more serious are the difficulties in getting the data on the institutional variables, especially on their diverse roles in the generation and transmission of development impacts. Talking about the data on institutions, it is important to note that since this study involves multiple institutions transcending sectoral boundaries and since sectoral institutions vary across provinces, it is essential to select the study area to be entirely within a single jurisdictional boundary. On this consideration, the evaluation context is confined to the North Central Province, which accounts for about 80%of the basin area.

8.6.1 Stakeholder Perceptional as Data Source

Lack or absence of data on most variables does not, however, mean a complete absence of information on institutional variables and their roles in development implementation because such information are constantly processed and stored in people involved in the development process either as planners and implementers or as beneficiaries. In view of this fact, a carefully conducted perception survey can unearth the untapped but highly relevant information that the individuals and society use regularly in making various decisions. Such information embodied in individuals is particularly valuable for the analysis of institutional roles and development synergy because they have many desirable properties often missed in observed data. For example, unlike observed data characterizing a past and static situation, the perception data can capture and synthesize objective, subjective, and aspiration-related considerations. It is also theoretically legitimate in view of the subjective nature of institutions (Commons, 1934; V. Ostrom, 1980; Douglas, 1986; E. Ostrom, 1990) and the role that the 'subjective model' of the 'agents of institutional change' plays in institutional change and performance (North, 1990). As a result, there is a long tradition of using such data for institutional analysis (e.g., Knack and Keefer, 1986; Gray and Kaufmann, 1998; Barret and Graddy, 2000; Kaufmann, Kraay, and Mastruzzi, 2006). Qualitative data are also used even in cases such as impact assessment (Neubert, 2000; Coudouel, Dani, and Paternostro, 2006).

Perception can be used as an evaluation mechanism not only to synthesize variables of different domains and conception but also to operationalize 'adaptive instrumental evaluation', where the outcomes are evaluated in positive and relative terms with respect to reference points that are not static but change with learning and expectations (Tool, 1977; Kahneman and Tversky, 1984; Bromley, 1985). In view of this property, perception-based information is similar in format and quality to those derived from alternative non-market data generation techniques such as 'Delphi', 'Contingent Valuation', and 'Stated Preference' (Saleth and Dinar, 2004). It is on the strength of these theoretical and practical considerations that this paper uses the stakeholder-based *ex-ante* qualitative information as a basis for the empirical evaluation of the model of institution-impact interactions.

Understandably, the empirical approach used in this paper is underpinned by two inter-related facts: (a) practically valuable information on the status and performance of institutions and on spread and intensity of development impacts are constantly processed, updated, coded, and used in various forms and in many real world situations of impact assessment and decision-making and (b) such real but latent information can be unearthed and captured with an innovative procedure that explicitly recognizes the central role of stakeholders both as change agents and as information source for the evaluation of institutional impacts and development synergies. Thus, the two key components of the empirical approach are the selection of a suitable sample of stakeholders and the elicitation of their perception-based information on all the variables needed for estimating the structural model.

The sample of stakeholders selected for data collection includes 67 persons, who are directly involved in development planning, implementation, and evaluation in the Kala Oya Basin. The sample covers government officials at different levels (32), researchers/academics (32), and farmers/community leaders (3).[9] For collecting the information on all the 32 variables included in the model specified in equations 1–21, a survey instrument was developed and administered to the sample stakeholders (Saleth et al., 2006). The data was collected during field visits in the Basin and the cities of Colombo and Kandy during May 2006. The descriptive statistics for the 32 variables are presented in Table 8.2.

8.6.2 Model Results and Institution-Impact Analysis

Assuming a simple linear form for all the equations and using the stakeholder-based qualitative information, we estimated two versions of the model of institution-impact interaction. The first is a single equation model, where food security is postulated as a simple linear function of all the remaining 31 development, impact, and institutional variables. This simple model captures the conventional approach, which

[9] Notably, these stakeholders, though knowledgeable of the region and its development process, are not all necessarily from the study region or the direct beneficiaries of the development interventions. This is partly to avoid the potential bias and partly to address the macro-micro dichotomy evident in empirical impact evaluation literature, i.e., micro evaluations report considerable impact whereas macro evaluations find little or no impact, or *vice versa* (Neubert, 2000; Coudouel, Dani, and Paternostro, 2006).

Table 8.2 Remaining variables

BULKWATD $= f_1$ (*SYSREHAB*) [1]
CROPDIVR $= f_2$ (BULKWATD) [2]
CROPATEN $= f_3$ (*CUSINSTN*, CROPDIVR) [3]
WATINSTN $= f_4$ (*CUSINSTN*, BULKWATD) [4]
LANPRODY $= f_5$ (*LANTENUR*, CROPATEN, LANHELTH, FAMINCOM, *FAMINSTN*, *SYSREHAB*) [5]
WATPRODY $= f_6$ (CROPATEN, LANPRODY, *WATINSTN*, BULKWATD, *SYSREHAB*) [6]
LANHELTH $= f_7$ (CROPATEN, WATPRODY, *SYSREHAB*, BULKWATD) [7]
FODPRODN $= f_8$ (CROPATEN, LANPRODY, *CUSINSTN*) [8]
NFAMENTS $= f_9$ (CROPATEN, *RDVPOLCY*) [9]
LABPRODY $= f_{10}$ (LANPRODY, RURALEMP, *WAGELAWS*) [10]
WAGERATE $= f_{11}$ (LABPRODY, *WAGELAWS*) [11]
RURALEMP $= f_{12}$ (LANPRODY, WAGERATE, NFAMENTS, LIVSTOCK *WAGELAWS*) [12]
CULTCOST $= f_{13}$ (*LANTENUR*, CROPATEN, *FAMINSTN*, RURALEMP, *WAGERATE*, *SUBPOLCY*) [13]
AGLINCOM $= f_{14}$ (LANPRODY, *MRKINSTN*) [14]
FEDSUPLY $= f_{15}$ (CROPATEN, *CUSINSTN*) [15]
LIVSTOCK $= f_{16}$ (FEDSUPLY, *TRDPOLCY*) [16]
FODAVAIL $= f_{17}$ (FODPRODN, *SAMPOLCY, MKTINSTN*) [17]
FODPRICE $= f_{18}$ (FODPRODN, *PRICREGL, MKTINSTN*) [18]
FAMINCOM $= f_{19}$ (AGLINCOM, CULTCOST, LIVSTOCK, *SUBPOLCY*) [19]
LABINCOM $= f_{20}$ (RURALEMP, *WAGERATE*, LIVSTOCK, *SAMPOLCY*) [20]
FODSECTY $= f_{21}$ (FODAVAIL, FODPRICE, FAMINCOM, LABINCOM, LIVSTOCK) [21]

assumes away the specifics and dynamics of interaction. The second version is the system model, which specifically captures the mechanics of impact generation and transmission in terms of 21 equations linked both sequentially and simultaneously.

The single equation model was estimated using the Ordinary Least Square (OLS) procedure whereas the system model was estimated using the Three-Stage Least Squares (3-SLS) procedure. Although the OLS results of the single equation model, which captures the conventional approach to institution–impact interaction, are not provided here, the key point to report here is the fact that none of the institutional variables were significant as are the variables representing the three development interventions. Even among the 17 impact variables, only five are significant at the level of 20% or better. These significant impact variables are: LABPRODY, WAGERATE, AGLINCOM, FAMINCOM, and LABINCOM. Notably, all of them, except AGLINCOM, have the expected positive effect. The negative effect of AGLINCOM, especially given the positive effect of FAMINCOM, is clearly inconsistent with expectation, as it suggests a negative association between agricultural income and food security.[10]

[10] This inconsistency taken with the insignificance of institutional and development variables clearly suggests the potential for serious anomalies when a single equation model is forced to a condition with a complex set of sequential and simultaneous interactions among the model variables. Such problems are also unavoidable when the roles of institutions are treated superficially or exogenously against the reality of their intricate and endogenous role within development process.

Table 8.3 Descriptive statistics

Variables	Mean	Standard Deviation	Minimum	Maximum
FODSECTY	5.07	1.59	0.75	8.00
CROPDIVR	6.04	1.79	2.00	10.00
SYSREHAB	6.75	1.19	1.67	8.83
BULKWATD	6.32	1.75	1.00	9.00
CROPATEN	5.60	1.00	2.79	7.57
LANPRODY	6.84	1.40	2.63	10.00
WATPRODY	7.29	1.42	4.00	10.00
LABPRODY	4.94	2.21	1.00	9.00
RURALEMP	5.31	2.08	1.00	10.00
WAGERATE	6.10	1.27	2.50	8.50
CULTCOST	5.66	1.68	1.00	8.00
AGLINCOM	6.90	1.49	3.00	10.00
LANHELTH	7.62	1.33	3.50	10.00
FOODPROD	5.22	1.23	2.33	7.67
NFAMENTS	7.07	1.29	2.25	9.50
FEDSUPLY	5.32	1.43	1.00	8.00
LIVSTOCK	3.64	1.62	0.90	7.90
FAMINCOM	5.50	1.09	3.00	9.00
LABINCOM	4.64	1.31	2.00	8.00
FOODAVAL	5.24	1.36	2.50	8.50
FOODPRIC	4.37	1.31	1.50	7.50
LANTENUR	6.20	1.15	3.56	8.33
WATINSTN	5.03	1.88	1.00	9.00
FAMINSTN	5.52	1.68	1.00	9.00
CUSINSTN	4.71	1.28	1.40	7.60
RDVPOLCY	5.07	1.85	1.50	9.00
MKTINSTN	5.10	1.35	1.67	9.33
WAGELAWS	3.51	1.74	1.00	8.50
TRDPOLCY	6.57	1.41	3.00	9.00
PRICREGL	4.62	1.57	1.00	8.75
SUBPOLCY	6.82	1.38	3.00	10.00
SAMPOLCY	5.12	1.97	1.00	10.00

In stark contrast, the system model results presented in Table 8.3 clearly demonstrate the policy insights that can be derived with a more realistic treatment of institutions, especially considering their mediating roles both in the generation and transmission of development impacts. Development impacts and institutions influence each other. The key aspect to note from Table 8.3 is the way both the institutional influence and the development impact are transmitted across the equations. The operational mechanisms for such transmissions are obviously the sequential and simultaneous interactions that occur among the development, institutional, and impact variables. Considering this, our interpretation of the results will proceed along the equations, to show how the dependent variables in the intermediate equations capture and transmit the development and institutional impacts finally onto the ultimate development goal of food security. In the process, we will also show the relative magnitude and statistical significance of different institutional and impact variables and indicate possible weak spots and missing links within impact pathways.

The System as a whole explains 68 percent of the variation in the independent variables. To begin with, the result for Equation [1], which postulates the relationships between two development interventions: SYSREHAB and BULKWATD, shows that the intervention related to water infrastructural improvement has a statistically significant positive effect on the intervention aimed at improving the institutional dimension of water distribution. The results provide evidence for development synergy and for the influence of development on institutional performance. The same can also be seen in Equation [2], where BULKWATD has a statistically significant negative effect on CROPDIVR, suggesting that the institution-related development intervention of bulk water distribution tends to reduce the prospects for crop diversification.[11] Apart from the infrastructural and institutional constraints, there are also other difficulties, especially those emerging from customary tendencies in crop choice. The results for Equation [3] clearly show that customary institutions (CUSINSTN) are more powerful than the economic and technical prospects for diversification (CROPDIVR) in determining the crop pattern (CROPATEN). Even though BULKWATD has not promoted crop diversification, it has a strong positive effect on water institutions, especially by strengthening farmer associations and promoting better water distribution. This is clear from the results of Equation [4].

Equation [5] provides statistical evidence for the relative role of physical, agronomic, and institutional factors in determining land productivity. As the results show, although the soil fertility and land health (LANHELTH) is the most dominant factor, institutional factors such as the FAMINSTN covering the extension and input supply systems and LANTENUR covering the tenure security are also important in influencing land productivity. Notably, the development intervention of SYSREHAB has direct negative effect on land productivity, though, as we will see in equation [7], it has an indirect but statistically more significant positive effect via LANHELTH.[12] Equation [6] shows that water productivity is influenced positively by land productivity whereas only negatively by the crop pattern.[13] While water

[11] The result here is not surprising because the policy of providing bulk water to farmer groups has not yet solved the issue of volumetric allocation to individual farmers, which is essential for independent crop decisions. Under the current conditions of canal infrastructure and water release policies, the introduction of volumetric allocation favorable for crop diversification remains a major practical challenge.

[12] This is an important aspect of development impacts, as simple and one-dimensional approaches to impact assessment may miss not only the relative magnitude and directions of impacts transmitted in multiple channels and pathways but also the role and influence of institutions, which are operating across these channels and pathways.

[13] This negative effect is understandable in view of the crop pattern in the region being dominated by food crops, which, as explained in the context of previous equations, is due to strong role of customs in crop choice and the poor prospect for crop diversification.

institution is not at all significant as a determinant of water productivity, bulk water policy has a significant positive effect on the same. Equation [7] shows that of the four variables, SYSREHAB and CROPATEN are both significant and with the expected positive impacts on LANHELTH. But, the behavior of the other variables (BULKWATD and WATPRODY) seems to be spurious.[14]

Turning to Equation [8], the variables having the significant positive impacts on FOODPROD are the CROPATEN and LANPRODY. It shows clearly that the level of food production is determined by the food crop-dominated crop pattern as well as the productivity of land, which actually captures the positive impacts of the institutional and physical variables such as land security, farm institutions, and soil fertility, as we have seen in Equation [5]. Interestingly, as can be seen from Equation [9], even though the crop pattern is dominated by food crops, it has a significant positive effect on the prospects for non-farm enterprises. This is partly due to the fact that most non-farm activities observed in the region are linked to the processing and marketing of food crops, especially paddy. But, active rural development policy is also contributing to the growth and diversification of rural non-farm activities. Equation [10] shows that the level of labor productivity is determined not by land productivity but primarily by the level of rural employment and by the wage rate and working conditions as influenced by the prevailing rural wage laws and regulations in the region. This is not surprising because with the similar crop pattern and productivity levels, land productivity, unlike the other factors, may not explain well the variations in labor productivity. This perspective is reinforced actually by Equation [11], where labor productivity and rural wage laws are the dominant factors determining the wage rates. It can also be noted that NFAMENTS, the variable capturing non-farm prospects, is not at all a significant factor in effective farm wage rates. This is, in part, due to the weak status of non-farm activities and no flow of workers between farm and non-farm sectors.

In the case of Equations [12], none of the factors postulated to affect cultivation costs are significant, even though the wage rates and the subsidy policy are relatively more important and have the expected positive and negative effect respectively. As per the results for Equation [13], CROPATEN and CUSINSTN have a positive and statistically significant effect on the potential for feed supply (FEDSUPLY). The domination of food crops, especially paddy, contributes to feed supply in terms of crop residues whereas customary institutions add to the same in terms of feed supply in terms of open grazing and biomass collection. But, as can be seen from the results for Equation [14], feed supply potential may not necessarily be the only factor to explain the prospects for livestock development (LIVSTOCK). As a result, livestock development can be lower even with a higher potential for feed and fodder supply, implying a negative association between the two.[15] Coming to Equation [15],

[14] Since BULKWATD has a significant negative effect on LANHELTH whereas WATPRODY is not at all significant, there seems to be the problem of multicollinearity among them. In fact, as we have seen in Equation [6], there is a strong positive association among them.

[15] The undeveloped potential of livestock development and the poor utilization of available feed/fodder observed in the study region are clearly consistent with the result obtained here.

farmer's income (FAMINCOM), which covers income only from farm operations, is influenced positively by both land productivity and cultivation cost.[16]

In equation [16], only three of the five variables postulated to affect rural employment are statistically significant. Of these, land productivity has a positive effect, but wage rates and livestock development have a negative effect. The inverse association between wage rates and rural employment also implies both the higher wage rates and labor scarcity.[17] Notably, *Samurthi*, the government's poverty alleviation program, does not have much effect on food availability, though, as we will see in Equation [20], it does have a significant role in augmenting labor income. Equation [17] model the association that food price has with food production as well as with the two institutions representing respectively the price regulations and market system. The results show that all the three variables are significant with a positive effect suggesting, thereby, that food prices continue to rise despite increasing production, procurement-related price regulations, and market expansion. This means that market and regulation mechanisms are not effective in moderating food prices.

Equations [19] and [20] evaluate the relative size and direction of the effects of factors determining respectively the farm and labor incomes. Although the income from agricultural operations has the dominant positive effect on farm income, the same from livestock remain also to be significant. Interestingly, cultivation cost, which has a positive effect on agricultural income in Equation [15], now has a significant negative effect on farm income. Among the significant factors affecting labor income, livestock remains dominant, though the government's poverty alleviation program of *Samurthi* also has an important effect. Coming now to Equation [21], this is the ultimate equation in the system, as it captures the various direct and indirect effects of development, impact, and institutional variables that flow through the intermediate equations. The results of this equation are very interesting because it shows that food prices rather than its availability that is more important. This means that from the perspective of promoting food security, the factors affecting food prices such as food production and the associated institutions in production, marketing, and distribution are very important. Similarly, the result that farm income has the positive and dominant effect on food security as compared to labor income suggests that food security is stronger among people with access to land than among those without that access. The insignificance of livestock variable suggests that the food security role of self-consumption from home grown livestock/ poultry products (e.g., milk, egg, and meat) is not at all important.

---

[16] The positive effect of CULTCOST, unlike that of LANPRODY, is somewhat unexpected, particularly given the prevailing concern in the study region about the income implications of rising cost of cultivation. However, the result suggests that agricultural income is rising in the face of increasing costs thanks to the possible neutralizing role of land productivity.

[17] The results for Equation [16] are intuitively consistent as they show that food availability in the market is positively influenced by food production on the one side and distribution-related market institutions on the other side. However, it can be seen that the production side plays a relatively stronger and more dominant role.

Table 8.4 System model of institution-impact interaction

		3-SLS Results (System R^2 = 0.685)[(a)]			
Equation Number	Dependent Variable	Independent Variables	Estimated Coefficient[(b)]	T-Ratio	Elasticity at Means[(c)]
[1]	BULKWATD	Constant	0.754	1.053	0.120
		SYSREHAB	**0.824**	8.002	0.881
[2]	CROPDIVR	Constant	**7.690**	7.186	1.272
		BULKWATD	***−0.261***	−1.574	−0.272
[3]	CROPATEN	Constant	**3.621**	4.085	0.646
		CUSINSTN	**0.362**	4.482	0.305
		CROPDIVR	0.045	0.433	0.049
[4]	WATINSTN	Constant	−0.352	−0.341	−0.070
		CUSINSTN	−0.032	−0.229	−0.030
		BULKWATD	**0.876**	6.006	1.100
[5]	LANPRODY	Constant	−0.625	−0.349	−0.091
		LANTENUR	***0.154***	1.491	0.139
		CROPATEN	0.263	0.912	0.215
		LANHELTH	**0.721**	2.863	0.802
		FAMINSTN	**0.150**	1.812	0.121
		SYSREHAB	***−0.189***	−1.496	−0.187
[6]	WATPRODY	Constant	−1.065	−0.506	−0.146
		CROPATEN	**−1.668**	−2.762	−1.282
		LANPRODY	**1.991**	4.213	1.869
		WATINSTN	−0.291	−1.063	−0.201
		BULKWATD	**0.877**	2.611	0.760
[7]	LANHELTH	Constant	**2.861**	1.853	0.375
		CROPATEN	**1.146**	3.748	0.843
		WATPRODY	−0.220	−0.769	−0.210
		SYSREHAB	**1.041**	3.150	0.923
		BULKWATD	**−1.123**	−4.315	−0.930
[8]	FOODPROD	Constant	**−1.628**	−1.922	−0.312
		CROPATEN	**0.892**	4.844	0.958
		LANPRODY	**0.276**	1.938	0.361
		CUSINSTN	−0.008	−0.106	−0.007
[9]	NFAMENTS	Constant	0.115	0.085	0.016
		CROPATEN	**1.098**	5.049	0.870
		RDVPOLCY	**0.159**	2.262	0.114
[10]	LABPRODY	Constant	**−0.749**	−0.355	−0.152
		LANPRODY	0.378	1.090	0.523
		RURALEMP	**0.429**	2.205	0.462
		WAGELAWS	***0.234***	1.894	0.167
[11]	WAGERATE	Constant	**4.621**	3.497	0.757
		LABPRODY	**0.256**	2.110	0.207
		NFAMENTS	−0.032	−0.169	−0.037
		WAGELAWS	***0.126***	1.572	0.073
[12]	CULTCOST	Constant	**2.977**	1.646	0.526
		LANTENUR	0.107	0.456	0.117
		CROPATEN	−0.234	−0.459	−0.231
		FAMINSTN	0.015	0.092	0.015

(continued)

Table 8.4 (continued)

		WAGERATE	0.714	1.081	0.770
		SUBPOLCY	–0.163	–1.025	–0.197
[13]	FEDSUPLY	Constant	1.221	0.971	0.229
		CROPATEN	**0.444**	1.609	0.468
		CUSINSTN	**0.343**	2.420	0.303
[14]	LIVSTOCK	Constant	**8.702**	8.283	2.401
		FEDSUPLY	**–0.916**	–4.553	–1.346
		TRDPOLCY	–0.030	–0.282	–0.055
[15]	AGLINCOM	Constant	0.656	0.438	0.095
		LANPRODY	**0.522**	2.092	0.518
		CULTCOST	**0.366**	2.204	0.301
		MKTINSTN	0.117	0.939	0.086
[16]	RURALEMP	Constant	***10.912***	1.392	2.054
		LANPRODY	**1.049**	1.850	1.351
		WAGERATE	**–1.363**	–2.130	–1.566
		NFAMENTS	0.034	0.057	0.046
		LIVSTOCK	**–1.306**	–2.158	–0.891
		WAGELAWS	0.009	0.054	0.006
[17]	FOODAVAL	Constant	–0.316	–0.275	–0.060
		FOODPROD	**0.908**	3.586	0.904
		SAMPOLCY	–0.031	–0.407	–0.031
		MKTINSTN	***0.192***	1.574	0.187
[18]	FOODPRIC	Constant	–1.166	–1.100	–0.267
		FOODPROD	**0.697**	3.096	0.831
		PRICREGL	**0.154**	2.070	0.163
		MKTINSTN	**0.234**	2.197	0.273
[19]	FAMINCOM	Constant	**2.441**	2.454	0.444
		AGLINCOM	**0.560**	5.196	0.702
		CULTCOST	**–0.265**	–3.180	–0.273
		LIVSTOCK	***0.142***	1.468	0.094
		SUBPOLCY	0.027	0.523	0.033
[20]	LABINCOM	Constant	–2.398	–0.664	–0.517
		RURALEMP	0.125	0.831	0.143
		WAGERATE	0.301	0.953	0.396
		LIVSTOCK	**1.043**	3.006	0.815
		SAMPOLCY	**0.148**	2.107	0.163
[21]	FODSECTY	Constant	***3.688***	1.371	0.728
		FOODAVAL	0.244	0.809	0.253
		FOODPRIC	**–0.924**	–2.845	–0.798
		FAMINCOM	***0.755***	1.517	0.819
		LABINCOM	0.026	0.067	0.024

Notes: [a] System R2 for 3-SLS, which captures the explanatory power of the whole model.

[b] Bold coefficients are significant at 10 percent or better. Bold and italicized coefficients are significant at 11-20 percent.

[c] Elasticity at means are the weighted coefficients with the weights being the ratio of the means of the concerned dependent and independent variables, This standardization enables a comparison of the relative importance of the independent variables both within and across equations.

8.7 Conclusions and Implications

This paper has shown that an insufficient treatment of the impact enhancing role of institutions leads to substantial welfare loss whereas the ignorance of the impact synergies among past, ongoing, and planned interventions leads to biased impact assessment. Understandably, these problems have far reaching implications, especially for meta-development goals such as MDGs, which require effective institutions and an integrated approach to development planning and implementation. To help address these serious problems, this paper has developed an analytical framework and evaluation methodology and illustrated them in the empirical context of the Kala Oya Basin in Sri Lanka, using stakeholder-based qualitative information.

The analytics of the institution–impact–impact framework allows showing both the specific point at which different institutions influence the impact generation and transmission process as well as the mechanics of impact synergies among the past, ongoing, and planned interventions. The mathematical representation of this framework provides additional insights on the functional relations among the development, impact, and institutional variables and sequential linkages among the impact pathways. For policy purpose, a better understanding of all these analytics, mechanics, and linkages are all very valuable because they can help package and sequence interventions, and identify and strengthen the major impact transmission paths and their underlying institutions.

Despite the preliminary nature of the model specified and qualitative nature of the information, the results do provide considerable insights on the roles that institutions play in the generation and transmission of impacts across impact pathways as well as the impact synergies that a given development intervention derives from others. These synergies, in fact, make the institutional evaluation more complex but rich because they provide the scope for considering institutional and developmental linkages together within the process of development. Since the regression results are, in effect, the statistical representation of the consensus prevalent among stakeholders, there is ample support for most of the relations postulated by the system model of institution–impact–impact interaction. Since the system model unbundles the impact process and deciphers its transmission channels, it is able both to capture the flow and interactions of development impacts and to show which institutions affect what channel. These are valuable information for policy design, institutional analysis, and impact assessment.

From the perspective of policy design, the results suggest that when planning an intervention in a given region, it is critical to consider the potential synergies possible from past, ongoing, and planned interventions. In our study region, for example, the implementation of system rehabilitation has considerably facilitated the performance of bulk water distribution and this positive synergy has also enhanced the prospects of crop diversification. The results also indicate that the synergy among the interventions can be enhanced with a fine-tuning of land, water, agriculture, and market-related legal provisions, policies, and organizations. Although the institutions covered within our framework are not exhaustive, the results show that among the institutions considered, those operating in the production and marketing spheres are relatively more important in terms of their role in channeling the impacts on the ultimate goal of food security. Specifically, since food prices and farm income are the most dominant factors affecting food security, all their intermediary variables and their underlying institutions (e.g., market, price regulation, land tenure, and credit and extension) are very important.

Besides the production-related farm institutions and distribution-related market institutions, there are also major influences from national level policies and laws such as those related to farm subsidy, rural industrialization, poverty alleviation, and wage rates and working conditions. At the same time, customary institutions related to cultivation practices and common grazing lands have significant effects on crop choice and livestock development. Notably, customary tendencies towards paddy cultivation, though a serious constraint for crop diversification, have a positive effect on the supply side of food security. To what extent changes in the performance of these rural institutions could affect the ultimate goal can, in fact, be evaluated in terms of chain functions capturing how a marginal change in any of the institutions leads to a series of changes within the equation systems and culminates finally on the marginal change on food security. Similarly, how impact synergies among development interventions contribute to the final goal can also be evaluated in terms of the marginal changes in one or more of the variables characterizing various impact chains.

While the methodology is intuitive and the results are insightful, we can also recognize some of the limitations and scope for further refinements, especially those related to the specification and structuring of the equations. For example, the insignificance of all the variables in Equation [12] makes it redundant and, hence, creates a gap in the system. Either this equation has to the re-specified or to be excluded from the system. Similarly, the unexpected signs in the case of some variables, insignificance of crucial variables in the some equations, and the inclusion of variables with strong association as independent variables in the same equation are problems that are to be avoided with a more refined set of equations. Unfortunately, these problems limit the scope for undertaking simulations and sensitivity analysis useful to demonstrate how the marginal changes in institutional and development variables lead to changes in the intermediate and ultimate dependent variables. But, considering the preliminary nature of the attempt on the analytical approach, mathematical modeling, and quantitative evaluation of this model, some of the analytical and econometric limitations are only natural to be expected at this stage. The paper still has succeeded to provide an empirical illustration of an analytical framework and evaluation methodology for dealing with two of the most serious issues related to the assessment of the progress of meta-development goals such as the MDGs.

References

Barrett, S. and Graddy, K. (2000). Freedom, growth, and the environment. *Environment and Development Economics*, 5: 433–456.

Bromley, D.W. (1985). Resources and Economic Development, *Journal of Economic Issues*, 19(September): 779–796.

Bromley, D.W. (1989). *Economic Interests and Institutions: The Conceptual Foundations of Public Policy*, New York: Basil Blackwell.
Coudouel, A., Dani, A.A., and Paternostro, S. (2006). Lessons from the implementation of poverty and social impact analysis of reforms. In Aline Coudouel, Dani, A.A., and Paternostro, S., eds., *Poverty and Social Impact Analysis of Reforms: Lessons and Examples from Implementation*, World Bank, Washington DC, pp. 1–28.
Commons, J.R. (1934). *Institutional Economics*, New York: Macmillan.
De Silva, S., Ariyaratne, R., and Abayawardana, S. (2006). Kala Oya Basin, Sri Lanka: A resource and development profile, IWMI Working Paper, Colombo, Sri Lanka.
Department of Census and Statistics (DCS) & UN World Food Program (WFP), (2005). *Vulnerability of GN Divisions to Food Insecurity, Anuradhapura District 2004*. Colombo, Sri Lanka.
Dinar, A., and Maria, S. R. (2005). Can Water Institution Be Cured? A Water Institutions Health Index, *Water Science and Technology: Water Supply*, 15(6):17–40, 2005.
Haines, A, and Cassels, A. (2004). Can the millennium development goals be attained?, *British Medical Journal*, 329: 394–397.
Johnston, J. (1984). *Economic Methods*, 3rd edn. New York, McGraw-Hill.
Just, R.E., Hueth, D. L. and Schmitz, A. (2004). *The Welfare Economics of Public Policy: A Practical Approach to Project and Policy Evaluation*, Cheltenham, UK: Edward Elgar.
Kahneman, D., and Tversky, A. (1984). Choices, Values, and Frames. *American Psychologist*, 39(4): 341–350.
Kaufmann, D., Aart, K., and Massimo, M. (2006). Governance Matters V: Governance Indicators for 1996–2005, Research Department, World Bank, Washington DC (Available at: *http://ssrn.com/abstract = 929549*)
Kennedy, P. (1987). *A Guide to Econometrics* (Second Edition), Cambridge, MA: MIT Press.
Neubert, S. (2000). *Social Impact Analysis of Poverty Alleviation Programmes and Projects*, Illford, Essex: Frank Cass.
Neubert, S. (2006). Impact Analysis in Development Cooperation and Approaches Adopted in Research and Practice, German Development Institute, Bonn (mimeo).
North, D. C. (1990). *Institutions, Institutional Change, and Economic Performance*, Cambridge, MA: Cambridge University Press.
Ostrom, E. (1990). *Governing the Commons: The Evolution of Institutions for Collective Action*, Cambridge: Cambridge University Press.
Ostrom, V. (1980). Artisanship and Artifact. *Public Administration Review*, 40(July–Aug): 309–317.
Sahn, D. E. and Stifel, D. C. (2002). *Progress Towards the Millenium Development Goals in Africa*, Working Paper, Department of Agricultural Economics, Cornell University, Ithaca, New York.
Saleth, R. M. and Ariel Dinar. (2004). *The Institutional Economics of Water: A Cross-Country Analysis of Institutions and Performance*, Cheltenham, UK: Edward Elgar.
Saleth, R. M., Dinar, A., Neubert, S., Manoharan, S., Kamaiah, B., Abayawardena, S., and Ranjith Ariyaratne,R. (2006). Institutional Evaluation and Impact Assessment with Multiple Interventions and Meta Development Goals: Methodology Development with Empirical Application. Draft in progress towards an IWMI Research Report).
Tool, M. R. (1977). A Social Value Theory in Neo-institutional Economics. *Journal of Economic Issues*, 11(December): 823–849.
UN (United Nations), (2006). *The Millennium Development Goals Report*, United Nations, New York.
UNDP (United Nations Development Programme), (2006). How to Guide MDG-based National Development Strategies, UNDP, New York (http://mdg-guide.undp.org/)
World Bank and IFPRI, Undated, *Agriculture and Achieving the Millenium Development Goals*, Report No. 32729-GLB, World Bank, Washington DC.

Chapter 9
Resource Pricing and Poverty Alleviation: The Case of Block Tariffs for Water in Beijing

Ben Groom, Xiaoying Liu, Tim Swanson, and Shiqiu Zhang

9.1 Introduction

In recent years the arid North East of China, and in particular Beijing, has suffered sporadic shortages of water. The causes of these events are manifold and like most manifestations of scarcity, water scarcity has important demand and supply side elements. Although on the supply side drought events have contributed to water shortages in the past few years, it is the nature of water demand that presents perhaps the most important determinant of water scarcity in Beijing. On the one hand, as a downstream user, surface water supplies to Beijing have been reduced by increased demands, largely from agriculture, in upstream areas of the Chao River (Hou, 2001). On the other hand, unprecedented economic growth and rural-to-urban migration in China as a whole means that urban water demand has increased both as the populous increases and as households become wealthier.

Water shortages entail considerable costs to residents and industry alike as emergency regulatory measures limit the quantities of water available. Residents face additional financial costs as alternative sources are sought and as a result of associated ill-health, while industry suffers the costs of lost production.[1] Furthermore, as surface water supply becomes perennially uncertain the reliance upon groundwater in Beijing has increased. Although groundwater is in principle renewable in recent years extraction rates have consistently exceeded recharge and the resource has been mined. Not only does the mining of groundwater highlight the temporary, unsustainable nature of this solution but it also signals another source of economic cost as water levels fall, extraction costs increase and land subsidence occurs (Hou, 2001). Perhaps the greatest cost of persistent water shortages, however, is the additional and increasingly costly investment in supply augmentation that they invariably lead to.

[1] For instance, in 2002, the reduced quantities of water available in Beijing meant that the operating pressure was only sufficient to provide water to ground floor residents (Xiaoying Liu, personal comment). The annual economic costs to industry of water shortages and drought events have been estimated to be in the region of RMB230 billion ($30 billion) in China as a whole (Hou and Hunter, 2002).

P. Koundouri (ed.), *Coping with Water Deficiency*, 213–237.

Beijing is no exception to this rule as evidenced by the progress of the $12 billion, 1,300 km 'middle route' of the South-North water transfer project, connecting the Yangtze River with Beijing and Tianjin.

Despite the costs associated with water shortages, their increased frequency in recent years and the massive investments that are underway as a result, industrial and residential water consumption has historically been priced at a level far below the full financial cost (Hou and Hunter, 2002). That is, in the water scarce regional economy of the North East of China, water consumption has been and remains effectively subsidised by the state. This situation is common in many countries, arid and humid alike, and such a pricing policy for household consumption is not without economic justification.

Along with education, clean water represents an archetypal 'merit' good since it conveys benefits, such as good health, beyond the private consumer and to society as a whole. Combined with the fact that minimum amounts of water are necessary for survival the case for marginal cost pricing or indeed rationing water by price seems to fall apart when it is possible that certain households will be unable to afford water. Clearly this is a prospect most likely in poor households.[2] So, the subsidisation of water consumption to ensure the minimum consumption required for good health has powerful economic and social arguments based both on efficiency grounds: to ensure the wider social benefits, and equity grounds: to ensure access.

On the other hand, in the face of water shortages and increased scarcity, the price of water takes on another important role: a resource management instrument. Cast in this light there are strong efficiency and equity arguments for the introduction of marginal cost water pricing.[3] Firstly, water pricing will ensure that consumers make efficient water consumption decisions, both statically and dynamically. This will limit demand to efficient levels and reduce the pressure upon primary resources. Secondly, revenues are generated which can ensure the maintenance of reticulation systems and the reliability of, and hence access to, water supply (Sterner, 2003).

This combination of water scarcity and revenue insufficiency provided the principle motivation for the Pricing Law of the Peoples Republic of China (PLPRC), enacted in 1998, as it relates to water resources. This law, along with more recent sister legislation sets out a programme of in-depth restructuring of the water sector. Specifically, the associated water-pricing regulations, the most recent of which came into force on 1 January 2004, set out general principles for pricing including provision for full resource cost recovery and seasonal pricing. Indeed, in the aftermath of the PLPRC cities such as Beijing have undergone substantial increases in the price of water, doubling in real terms between 1998 and the present day. Despite these increases tariffs remain below the best estimates for the long-run marginal cost (LRMC) and,

[2] See, e.g., Rock and Seckler, 1997 for a discussion.

[3] Here marginal cost may constitute the costs of provision, resource rents and other opportunity or environmental costs. With this interpretation marginal cost pricing can induce efficient consumption of water.

given the costly supply-side augmentation that has occurred in tandem, at the time of writing water tariffs in Beijing will have more than double again to reflect the full financial cost of supply.

Water pricing is commonly perceived to be something of a panacea for water scarcity problems in arid economies, mainly due to its efficiency and revenue-generating properties, and there are countless examples where such a policy has been successful on both counts. However, for the reasons cited above, the pursuit of efficiency is frequently in tension with commonly held perceptions of fairness and equity, and often at loggerheads with the pursuit of poverty alleviation. Pricing policies for necessities such as water are frequently controversial and unpopular. In addition to the merit good arguments above, another obvious reason for this antipathy is that necessities such as water generally make up a larger proportion if incomes for low-income households, and hence the impacts of price increases which move towards full cost fall disproportionately on the poor. The sum of these arguments means that the political economy of water resource management is such that policy makers are often reluctant to introduce water pricing at all, let alone align water prices with their full financial or economic cost. Hence, the status quo is a common outcome in this policy arena. Once more, Beijing is no exception to this and in July of 2004 the implementation of increased water tariffs was postponed as a result of fears about the disproportionate impact upon the poor.[4]

One oft cited pricing policy which appears to offer a way out of this impasse and alleviate the equity and poverty concerns surrounding water pricing is the Increasing Block Tariff (IBT). IBTs charge all water consumers a lower, frequently subsidised rate for a 'lifeline' quantity or block of water, the amount of which usually reflects the minimum household requirements for good health. Higher 'penalty' rates, usually in excess of full financial cost, are levied for consumption over and above this lifeline block. So, while a uniform tariff, despite its efficiency qualities, may have profoundly negative income effects on precisely those parts of the population least able to bear them, the IBT system is often thought to alleviate these problems by shifting the financial burden from low water consumers to high. In this way the equity–efficiency argument appears to be circumvented.

For these reasons an IBT regime has been proposed for Beijing.[5] However, there are a number of reasons why it is not immediately clear that such a pricing policy will circumvent the trade-off. For example, this would not be the case if rich households systematically have fewer members than poor households and the lifeline block of the IBT is defined at the household level. Furthermore, the welfare effects

[4] A similar line on water pricing for residential users has been taken in the UK where a flat rate for water consumption remains (O'Riordan et al, 1998).

[5] Many previous studies have demonstrated this negative impact of water pricing on the poor populations within developing countries (see, e.g., Koundouri et al 2003), meanwhile in developed countries the regressive effect of various water pricing policies is also evident (Smith and Rajah 1993, Agthe and Billings 1980). So it almost inevitable that at some point an IBT regime would be proposed for Beijing.

of price changes depend upon the nature of the responses for individual households – price and income elasticities –, family size relative to the subsidised block and a whole host of other household level factors such as the ownership of consumer durables.

The uncertainty in the impact of IBTs in Beijing is the motivation of this paper. In what follows we undertake an analysis of residential water demand in Beijing using the available data from the Chinese Household Income and Expenditure Survey (HIES). We obtain estimates of the Price and Income Elasticities of Demand (PED and IED respectively) and the extent to which these differ across income groups. Using these estimates we then present an analysis of the welfare impacts on different income groups of two proposed water pricing policies in Beijing: a uniform pricing policy at the Long Run Marginal (financial) Cost (LRMC) for water and the IBT system. More precisely, we obtain exact (Hicksian) welfare estimates for different income groups of the impact of moving from an economically efficient uniform tariff to an IBT tariff structure which is usually assumed to be more equitable at the expense of efficiency (Sterner, 2003, Whittington and Boland, 2000). We show that the latter does indeed reduce the impact of water pricing upon the lowest quintile of income, improving their welfare by an amount equal to 2.7% of income. Ultimately however, these findings need to be considered in the light of the other practical implications of IBTs and the empirical strategy that is employed here.

9.2 Water Pricing in Beijing: Efficiency vs Equity

Historically, water consumption has been subsidised in Beijing. Table 9.1 shows the evolution of water prices for Beijing for the period 1984–2003. However, it is interesting to note that the majority of residential water consumers have been monitored with water meters since 1980, and the scope of metering has increased (Hou, 2001).[6] The real price of water has increased in recent years in line with the provisions of the PLPRC and sister legislation such as the Management Method for Pricing Urban Water Supply or National Guidelines on Water Tariffs (NGOWT), of September 1998, the Water Act of 2002, and 'Measures on Water Price Management of Water Conservancy Project' which took effect in 2004.[7] The eventual aim is for tariffs to reflect full cost recovery by 2008 and by current estimates the full financial cost of water in Beijing is 6RMB/m^3.

The merit good qualities of clean water, that is, the health benefits conferred beyond the private user to society as a whole, coupled with the fact that water is ultimately necessary for survival means that rationing water by volumetric pricing,

[6] It is widely documented that some households share water meters in Beijing, and that in certain cases one meter serves an entire residential block (Hou and Hunter 2002; Xiaoying Liu. personal communication).

[7] The State Council Circular No. 45.

Table 9.1 Water tariffs in Beijing 1984–+ (RMB/m^3)

Year	Water tariff	Waste tariff	Total	CPI	Real tariff	Real MC	Real subs
2003	2.3	0.6	2.9	–	–	–	–
2002	2.0	0.5	2.5	107.22	2.33	5.60	3.27
2001	1.6	0.4	2.0	103.10	1.94	4.85	2.91
2000	1.3	0.3	1.6	100.00	1.60	2.00	0.40
1999	1.0	0.1	1.1	96.62	1.14	1.37	0.23
1998	1.0	0.1	1.1	96.04	1.15	1.37	0.22
1997	0.5	0	0.5	93.79	0.53	1.41	0.88
1996	0.3	0	0.3	89.07	0.34	0.47	0.13
1995	0.3	0	0.3	79.81	0.38	0.53	0.15
1994	0.3	0	0.3	68.04	0.44	0.62	0.18
1993	0.3	0	0.3	54.48	0.55	0.77	0.22
1992	0.3	0	0.3	45.78	0.66	0.92	0.26
1991	0.12	0	0.12	41.65	0.29	1.01	0.72
1990	0.12	0	0.12	37.22	0.32	1.13	0.81
1989	0.12	0	0.12	35.32	0.34	0.55	0.21
1988	0.12	0	0.12	30.13	0.40	0.58	0.18
1987	0.12	0	0.12	25.03	0.48	0.60	0.12
1986	0.12	0	0.12	23.05	0.52	0.54	0.02
1985	0.12	0	0.12	21.58	0.56	0.43	–0.13
1984	0.12	0	0.12	18.36	0.65	–	–

such that water is consumed only by those who can afford it, has the potential to leave low-income households high and dry. Of course this is a particular concern in developing countries. The corollary of this is that the lack of access to water is clearly a significant dimension and cause of poverty and for this reason, access to clean water is the cornerstone of many poverty alleviation strategies. It is uncontroversial to state that provision needs to be made in social or economic policy to ensure that access to water is ensured.

It is in this sense that poverty, water and water policy are frequently difficult to disentangle and it is in light of powerful arguments like those above that opposition to water pricing and particularly volumetric pricing, is based[8]. In addition to this there is the fear that, even if willingness to pay is sufficient among poor households, that is, water is 'affordable', the burden of such expenditures falls disproportionately upon the poor. Again, there are good reasons for this assertion, since almost by definition expenditures on necessities like water make up a larger proportion of incomes for poor households[9]. Engel curves reflect the relationship between

[8] This argument holds for developed and developing countries alike. For example, in the UK volumetric water pricing has been argued against on this basis and a system of flat rates for water remain despite scarcity problems in certain regions (Herrington 1999).

[9] Remembering that the economic definition of a necessity is that its income elasticity of demand lies between zero and one.

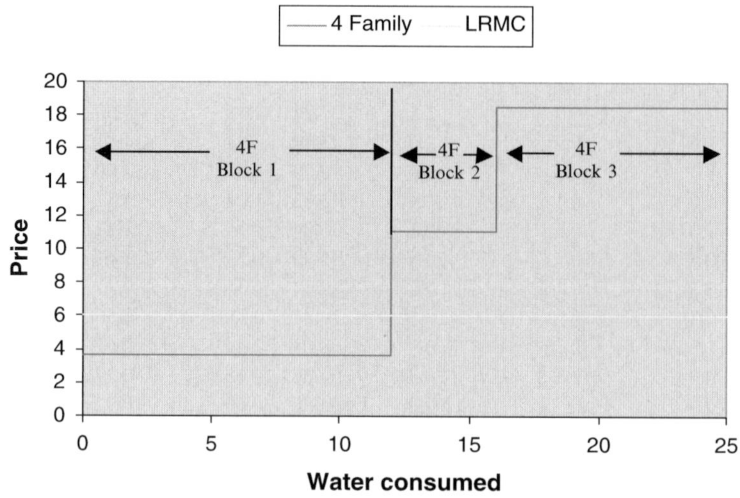

Fig. 9.1 The IBT and uniform tariff regimes proposed for Beijing (RMB/m^3)

income and expenditure for a particular good and the empirical Engel curve for water in Beijing reveals that expenditures upon water do indeed make up a larger proportion of income for poor households. At the subsidised prices of 2003, water expenditures made up 1.2% of incomes at the lowest income level and only 0.5% for highest. Figure 9.1 in Appendix 1 shows that this negative relationship is practically monotonic and linear in the log of income[10]. For this reason any change in the price of water is likely to increase the burden upon poor households disproportionately when compared to richer households. That is, water pricing is regressive, effectively worsens the distribution of disposable income and hence can be perceived as inequitable.

There are important equity arguments against volumetric marginal cost water pricing and these arguments ensure that the political economy of water pricing and water resource management is both complicated and engaging. On the one hand arid regions such as North East China suffer severe water scarcity and this acts as a significant constraint to sustained development. Water pricing has desirable efficiency properties in this respect. On the other hand, access to water is a dimension of poverty and water pricing raises important equity issues, despite its efficiency properties, and these factors are the source of significant opposition to water pricing. Indeed, opposition to water pricing has mounted in Beijing precisely as a result of equity concerns to the extent the in July 2004 the planned water price restructure

[10] The Engel curve was estimated using a semi-parametric regression technique. Figure A shows the outcome of the kernel regression.

was postponed[11]. In short there appears to be a classic equity–efficiency trade-off in this arena in which water policy and poverty alleviation are inextricably linked.

Clearly, there is a need for a policy instrument or instruments which can effectively ensure efficient use of water in a manner which is both equitable and contributes to, or at least does not detract from, poverty alleviation and equity issues. One such instrument attempts to circumvent this equity efficiency trade-off by combining a number of different features which address each of these concerns. Increasing Block Tariffs (IBTs) address the merit good qualities of water, issues of access and the unequal burden of expenditure on low income households by setting tariffs such that an initial 'lifeline' block is subsidised. The efficiency and scarcity issues are addressed by charging higher tariffs, usually above the cost of supply, for higher levels of water consumption. In combination, these features recognise that although water has merit good qualities which may justify subsidisation, these qualities tend to be limited to the minimal quantities required for good health and hygiene. Hence, the lifeline block is frequently determined by some estimate of this minimum quantity. Higher tariffs for consumption beyond this level reflect the scarcity of water and, desirably it is said, encourages more conservative use and allows a level of cross subsidisation to take place from high consumers to low[12].

These apparently attractive features have lead to IBTs being proposed for Beijing as an alternative to the uniform long-run marginal cost tariff of 6RMB/m^3. Figure (9.2) shows the structure of this tariff regime along with the uniform long-run

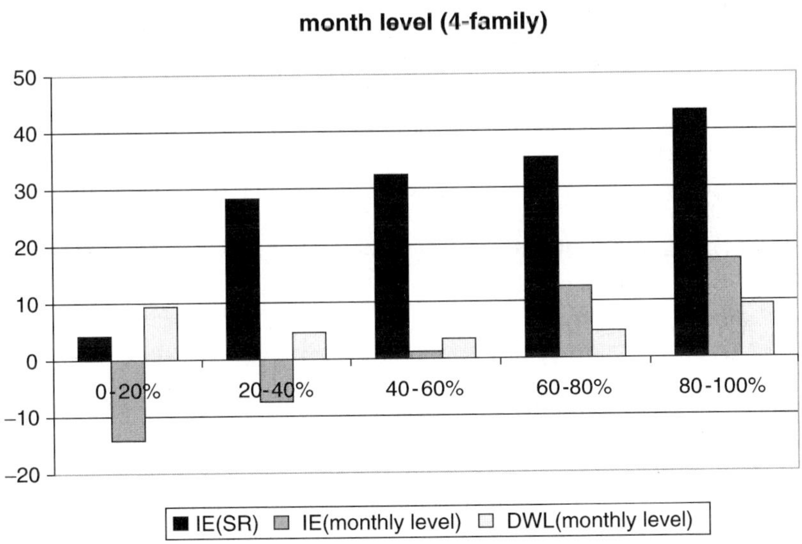

Fig. 9.2 Expenditure index and deadweight loss arising from a shift to uniform LRMC pricing to the IBT

[11] Transcripts of the public hearings regarding water pricing in Beijing are available on-line.

[12] For a complete discussion of the merits of IBTs see Whittington and Boland (2000).

marginal cost tariff. This shows that the proposed IBT structure is increasing steeply for household consumption in excess of the 'lifeline' block of 12 m ^{3}/month. The lifeline block has been defined on the basis of a four-member household with each member consuming 3 m^3 /month. The rationale for the four-family model is that 90% of households in Beijing have no more than four members, and hence the IBT structure aims not to constrain households on the basis of family size.

A number of questions remain with regard to the efficacy of IBTs in circumventing the equity–efficiency trade-off described above. It is not immediately clear that a tariff which is increasing in the quantity consumed will favour poor households. For example, if the lifeline block is based upon household size, as is the four-family IBT described above, then poor households are likely to face higher marginal tariffs, other things remaining equal. Similarly, the impact upon incomes and, more importantly, the precise welfare effects of IBTs will depend upon the nature of the response to price changes, that is, substitution and income effects. These factors are also likely to differ across income groups due to differences in preferences and the availability of substitutes (Koundouri et al., 2003).

Our objective in this paper is, therefore, to test the claims of that IBTs make with respect to equity and efficiency by estimating the exact (Hicksian) welfare effects for the movement from uniform LRMC pricing to the proposed IBT structure in Beijing. We do this by estimating the demand for water in Beijing for different income quintiles and hence estimating the responsiveness of water demand to price and income changes. Having done this we engage in a policy simulation to compare the two pricing regimes and the extent of the redistributive effect of IBTs. The results are discussed in light of the forgoing discussion and alternative policies to address the equity–efficiency trade-off described above. The next section outlines the empirical strategy and the two exact welfare measures we use for the purpose of our analysis.

9.3 Empirical Strategy

9.3.1 Demand Estimation

In order to obtain theoretically precise estimates of the welfare effects of alternative water pricing schedules for different income groups we follow Deaton and Muellbauer (1980) and Koundouri et al (2003) and estimate an integrable demand function for water which conforms to the theory of consumer behaviour. Such an estimation allows us to retrieve, with some restrictions, the parameters of the utility function for a representative consumer and hence allows us to obtain exact (Hicksian) welfare measures for price changes rather than Marshallian measures of consumer surplus found in other demands studies (e.g., Höglund, 1999). We undertake the analysis for income quintile groups in order to capture their different demand responses to price and income changes and we use the Almost Ideal Demand System (AIDS) of Deaton and Muellbauer (1980).

The demand curve for good *i* (e.g., water), in budget share form, that is implied by the AIDS model is:

$$w_{ih} = a_i + \sum_j \gamma_{ij} \ln p_j + \beta\left[\ln y_h - \ln a(p)\right] \tag{1}$$

where $a(p)$ is a function of a vector of prices *p* for all other goods which for estimation assumes a flexible functional form. Parameter γ_{ij} represents the own and cross price elasticities of demand for goods *i* and *j* (where $i, j = 1,\ldots n$). Appendix 1 provides an overview of the basic tenets of the AIDS model and shows the derivation of the demand system.

9.3.2 Welfare Measures for the Policy Simulation

Firstly, denoting the indirect utility function as $V(p, x)$, where *p* is the price vector and *x* is income, it is possible to define the utility for a household with income level x_0 under at the baseline prices p_0 as $V_0 = V(p_0, x_0)$. Following Koundouri et al (2003) it is possible to define the level of income $x_0 - \Delta$ which would leave the household with the baseline utility V_0 when faced with IBTs, reflected by the new price vector p_1 [13]:

$$V(p_1, x_0 - \Delta) = V(p_0, x_0) = V_0 \tag{2}$$

The quantity Δ is a measure of the compensating variation (Freeman, 1993, p53) and hence reflects the willingness to pay to obtain or the willingness to accept compensation for the change in prices from uniform tariffs to IBTs. This is our first exact welfare measure.

Secondly, we provide a measure of the deadweight loss: the difference in household utility that arises as a result of the price changes keeping expenditures at the baseline level. In short this is a measure of the extent to which consumers would prefer a revenue neutral lump sum subsidy (tax) as opposed to the commodity subsidy (tax). The deadweight loss is defined as:

$$DW = V(p, x^*) - V(p^*, x^*) \tag{3}$$

This represents our second exact welfare measure.

Armed with our estimates of demand and our measures of welfare changes we are able to answer our principle question: what are the welfare impacts of different pricing policies upon different income groups, specifically the poor?

[13] Appendix 2 shows how these measures are derived in the AIDS system.

9.4 Empirical Analysis

9.4.1 Data

The data is taken from the Chinese Urban Household Income and Expenditure Survey (HIES), the collection of which is undertaken annually by the State Statistical Bureau of China. The survey requires respondents to keep a daily expenditure diary for a full 12 month period. As with other surveys of household expenditure, the expenditure categories are numerous and detailed and include household durables as well as consumption items, incomes and numerous details concerning family members and characteristics of accommodation. Quantities are also recorded. Theoretically, the data is vetted for quality by an enumerator who visits each household once or twice a month.

The raw data is collected by the local State Statistical Bureaux and aggregate statistics are compiled and published on an annual basis in the Province Statistical review. In general, raw, household level, data is not available to the general public. One of the only ways in which the data is disaggregated in a publicly available form is by income quintiles. That is, data is presented as average expenditures for each of these 5, or occasionally more, income groups (Chinese Statistical Yearbook, 2003).

We used the HIES for Beijing in various levels of aggregation. The preliminary analysis was undertaken using annual published statistics over the period 1987–2002 which gave average expenditures and average household characteristics for income quintiles. The final analysis was undertaken using household level data collected directly from the Beijing State Statistical Bureau from the urban HIES. These data contain monthly information from 1,368 households in 9 districts of Beijing for the years 2002 to 2003[14]. Due to resampling there are only 645 households having a complete 24 months of observations. The survey provides the information about the district of each household, household's total expenditure and non-food expenditure in each month and also the utility expenditure, water quantity and water expenditure for each billing period. Besides these panel data at household level, we also have aggregated (quintile) data concerning average household characteristics such as possession of durable consumer goods, housing condition and expenditure on main foods. Finally, for the analysis of water demand, these data were combined with data on the real price of residential water (at 2000 prices) obtained from the Beijing Municipality shown in Table 9.1 above.

The household level data is deficient in two senses, reflecting the difficulties faced in obtaining household level data in Beijing. Firstly, there are very few variables

[14] The HIES surveys 1000 households every month. However, due to a degree of resampling, these data are an unbalanced panel, with 1368 households constituting the cross-section, and monthly observations for each household constituting the time-series.

available at the household level. It appears that the Beijing SSB were not accustomed to releasing the data at this level and therefore found it difficult to extract these raw data. Confidentiality was also reported to be an issue. Therefore, we have very few explanatory variables at the household level for household demand for water. Secondly, and equally important, there is only limited price variation during the period for which we have household level data, and this is true of the real tariff also[15]. In a city in which residents face the same marginal price for water, the only way to overcome this problem is to obtain data from additional time periods[16]. Unfortunately, these 2 years (2002 and 2003) were the only time periods available at the present time.

In light of the paucity of data at the household level we also made full use of the aggregated data obtained from the Chinese Statistical Yearbook for Beijing (henceforth, the time series data). These data are limited in their value since they obtain only information for broad groups: income quintiles, and hence are constrained in the extent to which they can provide generate statistical inference at the household level. However, these data are available for a longer time series and hence can be combined with the time series data for water prices in Beijing shown in Table 9.1. The following section describes the manner in which these sources of data were used in the estimation of water demand.

9.4.2 Estimation

The data limitations described above are not uncommon. On the one hand we have a long time series which provides variation in 'macro' variables, in our case the price of water, but very little variation in the 'micro' household level characteristics. On the other hand we have a cross section of households with variation in important micro characteristics but with very little variation in the important macro data. Initial estimations based solely on the aggregated data provided significant estimates for price elasticities, but not for income elasticities. The opposite was true for the household level data. In order to resolve this issue we used a two step method originally from Tobin (1950).

Firstly, we estimated the parameters relating to income elasticity of demand from the household level data: $\hat{\beta}$. Secondly, inserting these parameters into the budget share equation (1) we estimated the parameters which relate to price elasticity of demand using the aggregated time series data. In the both steps we use panel data techniques to estimate the following equation for water demand in budget share form based on equation (1) above:

[15] In fact there are only three different nominal prices for water during this 24 month period.

[16] Note that there was no spatial variation in tariffs in Beijing.

$$w_{ht} = \sum_k \rho_k \ln z_{ht} + \gamma_g \ln p_t + \beta_g \left[\ln x_{ht} - a \ln p_t - 0.5\gamma_g (\ln p_t)^2 \right]$$
$$+ \psi_1 SUMMER_t + \psi\, SARS_t + \psi_3 freq_h + u_h + e_{ht} \quad (4)$$

where $g = 1,..,5$ represents the income groups, t represents months and h represents the household. The variable *SUMMER* is a seasonal dummy variable, *SARS* represents the period of time during the SARS epidemic while *freq* represents the frequency with which the household recorded expenditures on water in a given year. z_{htk} represents the kth household characteristic. As is usual in such models f_h represents the time invariant household specific effect and u_{ht} represents the random error. In the second step we take the parameter estimates $\hat{\beta}_g$ as fixed and estimate the following equation using panel data methods on the data aggregated by quintile groups:

$$w_{gt} - \beta_g \ln x_{gt} = \alpha + \left(\gamma_g - \alpha\hat{\beta}_g\right) \ln p_t - 0.5\hat{\beta}_g \gamma_g (\ln p_t)^2 \quad (5)$$

$$+\psi_1 t + \sum_k \psi_{2k} \ln z_{kgt} + f_g + u_{gt} \quad (6)$$

In this way we were able to retrieve the parameters useful for the policy simulation with the data made available to us. In particular, PED and IED are calculated using the following formulae (Deaton and Muellbauer, 1980):

$$PED = \frac{1}{w_{ht}} \left[\gamma_g - \beta_g (\alpha - \gamma_g \ln p) \right] - 1 \quad (7)$$

$$IED = \frac{\beta_g}{w_{ht}} + 1 \quad (8)$$

9.4.3 Parameter Estimates

The approach outlined above yields estimates for PED which are statistically significant at the 1% level, of the expected sign and of a magnitude which is consistent with previous studies in this area (e.g., Billings and Agthe, 1980, Nauges and Thomas, 2001, Martínez-Espiñeira, 2002). The estimates for PED are on the high side compared to previous studies and closer to estimates for the long-run PED that appear in the literature (OECD, 1997, Nauges and Thomas, 2001)[17]. However, further analysis of the data suggests that the high estimates for PED may reflect the change

[17] Nauges and Thomas (2000) find long-run price elasticities of –0.4 in a panel study of French municipalities. Koundouri et al (2003) find estimates for PED ranging from -0.8 to -0.4 for residential water in Cyprus.

Table 9.2 Estimates of PED and IED for income quintiles[18]

Income groups					
Elasticity	0–20%	20–40%	40–60%	60–80%	80–100%
IED	0.135**	0.172**	0.106**	0.118**	−0.079**
PED	−0.619**	−0.617**	−0.536**	−0.583**	−0.565**

in the policy environment post 1998. A Chow test in the aggregated time series data revealed that there was a structural change in demand after 1998, that is after the PLPRC had been enacted responsiveness to price increased. In our analysis we chose to use the post 1997 time series data, since we felt that it better reflected the current consumer responsiveness and the active policy environment which was absent in prior to 1998. Indeed, the Chow test also revealed that demand for water was highly price inelastic in the period prior to 1997, a period during which the real price of water was low and frequently diminishing over time (see Table 9.1). The estimated parameters of interest are shown in Table 9.2.

Importantly, PED is higher for low income groups than it is for middle and high income groups. That low income groups are more responsive to price changes reflects the fact that water expenditure makes up a larger proportion of their incomes and as a result price changes are more rapidly internalised. It may also reflect access to outside sources for water, something for which we have no data. Having said this the differences between the groups are not great in practical terms: with a 10% rise in prices the low income groups will reduce consumption by 6.2% while the high income groups would respond with an approximate 5.6% reduction.

As one would expect, the estimates of the IED are between zero and one and hence water can be classified as a necessity in economic terms. The IEDs range from 0.17 for the 20%–40% quintile to approximately zero for the top income quintile. It should also be recognised that the income elasticity is higher for the low income households: they spend a higher proportion of additional income upon water than do high income households. Oddly, however, the IED for the highest quintile of income is negative, implying that water is in fact an inferior good for high income groups. Low or zero IEDs seem to be plausible since water consumption in Beijing is mainly for indoor use. Rich households have few private lawns which might be irrigated and private swimming pools are uncommon. In this sense water use is largely for its function as a necessary good for rich and poor alike. This in turn explains the low estimated budget elasticities for each income group[19].

[18] ** reflects statistical significance at the 1% level.

[19] We must be careful in our interpretation given the limitations of our data. For example, although we have corrected for the average family size in the income quintiles, there is very little variation on these average data, and ultimately little explanatory power. More importantly perhaps, there are very few explanatory variables and many of them are collinear in the regression. For example, data on average household purchase of consumer durables hides the discrete nature of such purchases at the household level. Taking an average of these purchases leads to collinearity of these variables.

The other determinants of the water demand were also important. The dummy variable *SUMMER*, as expected, has a positive effect on water use. Similarly the time trend variable *t* has a positive and significant coefficient, showing that there is a trend towards increased water consumption, which is not explained by income, price or other changes. Lastly, the water expenditure share decreases with the increase in variable *freq*, the frequency of payment. This suggests that the frequency of payments is an important determinant of overall consumption, with the frequency of payment potentially focussing households upon their consumption levels. Lastly, and as an indication of the limitations of the data, no household characteristics were significant, mainly because many of these measures were collinear and did not vary sufficiently.

9.5 Policy Simulation: Predicted Welfare Impact

9.5.1 Methodology

Using the parameter estimates and methodology outlined above, we compare pricing regimes and obtain exact welfare measurements for each income group. One of the practical problems that emerges in the process of this policy simulation is determining the level of consumption that households would choose when faced with a block tariff schedule. Although it is easy in principle to predict how demand would change with a change in the level of the uniform tariff, the non-linear budget set that households are faced with under a block tariff system makes predictions more difficult in general. However, the parameter estimates provided by the AIDS gives us almost all the information that we require in order to predict households' utility maximising choices. The issues surrounding IBTs and non-linear budget sets in general have been widely discussed and Appendix 3 illustrates the problem more clearly [20].

Consumers face a kinked budget constraint under IBTs, as illustrated by Fig. 9. B in Appendix 4. In order to predict where on this kinked budget constraint consumers will choose to consume water once the IBT has been introduced we propose the following procedure:

- *Predict conditional demands for water (equation (1)) for each block of the IBT using the parameter estimates and the different values of 'virtual' income and price associated with each block.*
- *If the predicted quantity for a particular block lies within levels that define the block then we know that we have found a tangent point like those in the Fig. 9. B. If the predicted quantity is not in the block then we know that this block is not*

---

[20] See for example Nordin (1975) Taylor (1976) and Moffit (1986) in the context of water resources.

a feasible choice for that household and the tangency point lies on the virtual budget constraint (the dotted line in Fig. 9.B).

- *Repeat for all blocks.*
- *If no tangent point can be found for a household assume that they consume* $\bar{w}$, that is, precisely at the kink[21].

By taking this approach we assign each household to one of the blocks of the proposed IBT in Beijing on the basis of their utility maximising behaviour.

9.5.2 Welfare Analysis

As described above, the Beijing Municipal Authority plans to implement an increasing block tariff in Beijing, despite the recent postponement. An alternative policy to this is to implement a higher uniform price reflecting the long run marginal cost (LRMC) of 6 RMB/m^3. Using the method described above we compare the IBT and uniform LRMC pricing regimes in terms of their relative welfare effects upon different income groups. We estimate the expenditure index and deadweight losses defined by the expressions (2) and (3). In each case we estimate the welfare impact with and without any demand response. We interpret these as the long run and short run responses resepctively[22]. Given our high absolute values for PED we feel that these two scenarios give reasonable bounds on the precise welfare effect for each income group. Fig. 9.2 shows the results of the policy analysis.

Fig. 9.2 shows the expenditure index and the deadweight loss, measured in RMB, for an average household in each income quintile resulting from a move from uniform pricing at the LRMC to the IBT proposed for Beijing: the 4 family tariff. The results take into account the seasonal variation in water demand, meaning that a household will consume in different blocks depending upon the month. The expenditure indices tell us the compensatory income that the households would require in order to have the same level of welfare under the IBT as under the uniform tariff. The black blocks represent the short-run impact upon the consumers and it can easily seen that all income groups will suffer a welfare loss in the short-run since this expenditure index is positive for all income groups. Clearly, however, the welfare loss per household is increasing with income and this shows that, in the absence of a demand response, the welfare impact of the IBT hits the rich households

[21] Note that consumers that have the potential to consume at either of the two kinks associated with the three block IBT proposed for Beijing are sorted according to the value of the utility function. Also, although the latter step may appear to be somewhat arbitrary, empirical evidence exists in the context of water consumption and more generally taxation to show that households do indeed cluster around kink points (Moffit 1996, Martinez Espineira 2002).

[22] We use these terms indicatively and do not claim that the long-run elasticities, which depend upon aspects such as the purchase of consumer durables, have been estimated here.

hardest in absolute terms. This reflects the fact that rich households find themselves in the highest blocks more often than the poor once the IBT is introduced and that poor households generally consume within the lowest block already under the uniform tariff. Indeed, 15% of households in the richest group fall in block 3, while only 2.5% in the poorest group. This measure of the welfare impact gives an extreme upper bound for the welfare simulation of IBT. Similarly, in relative terms the expenditure effects are progressive in the short-run, measuring 2.5% of the highest income groups income and mere 1% of the lowest income households.

The grey blocks in Fig. 9.2 show the long-run effects once the demand response is considered. This shows that poor households are actually better off under the IBT once the demand response has taken place since the amount of income that would be required in compensation is negative. That is, welfare for the poor households is improved under the IBT compared to the uniform tariff since water consumption in the first block is subsidised. The monetary value of the welfare increase is approximately RMB13 per month, equal to about 2.7% of monthly income for this group on average.

The same cannot be said for high income households however who, even after their response to the price increase, end up worse off than under the uniform tariff. Comparing this to the previous result it is easy to see that some of the burden of the cost recovery is shifted from the low income groups to high income groups under the IBT. Indeed, Fig. 9.2 shows that high income groups would require positive compensation to move from the uniform tariff to the IBT as shown by their positive long run index of expenditure. The welfare loss amounts to RMB18 per month, approximately 1% of income.

Another interesting feature of the policy simulation is to determine the distribution of consumers across the blocks of the IBT system. Table 9.3 shows this breakdown both before and after the demand response. It is noticeable that the demand response is significant with the majority of households consuming within the first block of IBT. This reflects a number aspects of our estimation. Firstly the relatively high estimated elasticities and secondly the extent to which the predictions for the policy analysis were significantly out of sample. It is worth remembering that the estimation was undertaken using price data which varied within an approximate range of RMB0.56–2.5, whereas we predict water demand at prices up to RMB18.5. Of course, by definition predictions will always be out of sample, however, the point to take from this discussion is that the credibility of this prediction needs to

Table 9.3 Water consumption after (before) the imposition of the 4 family IBT

Income quintile	Hshlds block 1	Hshlds block 2	Hshlds block 3	Total	Hshlds block 1 (%)	Hshlds block 2 (%)	Hshlds block 3 (%)
0–20%	122 (110)	0 (9)	0 (3)	122	100 (90.2)	0 (7.4)	0 (2.4)
20–40%	129 (105)	1 (11)	0 (14)	130	99.9 (80.8)	0 (8.5)	0 (10.7)
40–60%	140 (112)	3 (7)	0 (19)	143	97.9 (81.8)	0 (4.9)	0 (13.3)
60–80%	141 (106)	1 (15)	1 (16)	143	98.6 (78.3)	0.7 (10.5)	0 (11.2)
80–100%	141 (106)	6 (19)	0 (22)	147	95.7 (72.1)	4.3 (12.9)	0 (15.0)

be considered in light of the extent to which it is out of sample, and the parametric assumptions upon which it is based.

9.6 Discussion

Our analysis shows that the use of IBTs can reduce the impact upon the poor of water pricing policies in Beijing. The cross subsidy inherent in IBTs flows from high to low consumption households and this happens to coincide with a flow from richer to poorer households. Furthermore, as shown by Table 9.3, the water demand for high consumption households is stifled considerably by tariffs over and above the LRMC for the higher blocks. In this sense, water conservation is encouraged.

Does this circumvent the equity–efficiency trade-off that is omnipresent in the pricing of public utilities? At first glance this would appear to be the case: low income households benefit and richer high consumers are encouraged to be more conservative in their water use. However, these findings ought not to be seen in a vacuum and there exist a number of arguments against IBTs in the light of which our analysis ought to be cast.

Firstly, there is the issue of revenue sufficiency. As shown by Table 9.3, the elastic price response estimated means that the vast majority of consumers end up in the lowest block of the IBT. Hence most households will be subsidised in their water consumption. Such a prediction is important for the Beijing Tap-water and Sewage treatment companies since the tariff will fail to achieve cost recovery and hence the quality of the water supply (regularity, wastage, water quality etc.) will be hard to sustain or improve. Since altering the various facets of the IBT could resolve this problem (e.g., the size of the first block and the size of the steps), the obvious question is: what does the revenue neutral IBT structure look like? This question is beyond the scope of the current paper however.

Secondly, and more generally, some commentators suggest that IBTs introduce perverse incentives for water supply corporations/companies (Whittington and Boland, 2000, Sterner, 2003). Sterner (2003) notes that if the objective for the corporation is full cost recovery or for a private company to earn a return in capital, when an IBT is introduced the poor become a low priority, e.g., repairs since their consumption is generally subsidised. Ultimately, the tariff structure introduced specifically to assist the poor can perversely induce a lower quality service for poor households in the long run. Ultimately this is a question of regulation of public services, yet the potential for such incentives, and the regulation required to remove them, needs to be tallied against the potential welfare enhancing effects described above.

Thirdly, it should be made clear that the IBT system subsidises all water consumers, high and low, rich and poor, for the initial lifeline units of water consumption. At first glance this appears to tally well with the merit good qualities of water which presumably apply regardless of income levels. However, it is likely that high income households neither have particular problems with access to water nor need encouragement to consume water in quantities that fulfil these societal benefits. Hence, the economic justification for subsidies for all consumers, as exists under IBTs, is not entirely robust. Furthermore, as suggested by Whittington and Boland (2000) the full extent of the subsidy is only available when households consume the entire lifeline block. Not all poor households will do this of course and hence they will not receive the full subsidy.

Fourthly, there is the question of family size. It is frequently the case that poor households have larger families than richer households and although family size appears to be relatively uniform in Beijing, it is likely that at the lowest extremes of the income distribution households will lose out from a system based upon a four-member family.[23] This will be especially true where households share one metered connection.[24]

Finally, there is the question of efficiency. Despite the welfare gain for the poor, and the progressive nature of the tariff structure, there will always be an efficiency loss arising from the movement away from uniform pricing since not all consumers pay the marginal costs of water. Although the inefficiency introduced is quantifiable in theory this remains the focus of future work in this area. In sum these issues suggest that even with the implementation of IBTs the equity–efficiency trade-off involved in water pricing remains.

Alternative pricing and water management strategies exist which address the equity–efficiency trade. In Chile for example, the lifeline tariff was replaced by a means tested direct cash subsidy in certain urban areas. All households are charged the full cost recovery tariff but a refund was made available to households upon application and was conditioned upon the water bill being more than 15% of household income. Such a policy would tally well with the preferences estimated in Beijing which showed the presence of significant deadweight losses associated with the IBT and hence the efficiency improvement that could be achieved by using lump sum subsidies as opposed to water price distortion. Furthermore, by charging fractionally above the full cost recovery tariff addresses issues of revenue sufficiency and efficient use of water.

It must not be forgotten that the redistributive effects of water pricing policies are limited in the extent to which they can alleviate. There is a strong argument for untangling poverty, water pricing and water management issues and using general taxation to redistribute incomes while ensuring that water tariffs reflect the economic costs and ensure efficient water use.

[23] According to the aggregated data, the average family size for the highest quintile almost 3 while the average family size for the lowest quintile is almost 4.

[24] It is sometimes thought that IBT systems are more administratively costly for the water supply company while being confusing for a consumer, although the authors cannot find a concrete reference for this. Whittington and Boland (2000) provide an interesting discussion of these issues.

9.7 Conclusion

Water pricing provides yet another arena in which we are faced with equity–efficiency trade-offs. Although economists have historically been interested in both sides of this trade-off, in the eyes of many their policy prescriptions have erred towards the latter. The pricing of public utilities in developing countries provides perhaps the most pertinent and difficult backdrop against which to discuss these issues since, to a large extent the provision of these utilities and the distribution of resources has been neither efficient nor equitable. Furthermore, since water is a necessity and has the characteristics of a merit good there remain clear and convincing arguments for its provision to be subsidised. That the lack of access to clean water is a significant dimension of poverty generates a tension in arid countries between poverty alleviation and efficient resource management since water pricing takes on the role of a resource management instrument and yet tends to impose disproportionate costs on the poor. Therein lies the crux of the trade-off.

In this paper we have analysed one pricing policy which, it is often claimed, is able to circumvent this trade-off and generate efficient consumption decisions at the same time as maintaining affordable access to necessary public goods such as water and electricity. Increasing Block Tariffs (IBTs) appear to achieve this goal by providing 'lifeline' units of water at subsidised rates and any consumption in excess of this is charged at close to full cost. The supposed positive equity aspects of IBTs are predicated upon a close relation between water consumption and income, that is, most low income households will remain in the subsidised block.

We have shown in the context of Beijing that IBTs do indeed soften the blow to the poor of increasing water prices, and the subsidy they receive represents 2.7% of income. Furthermore, since water demand appears to be relatively price elastic, the proposed IBT reduces household water demand significantly. However, despite being more equitable, our analysis shows that the proposed IBT for Beijing may not be revenue sufficient since most households consume within the lifeline block. This raises questions about what a revenue neutral the structure of the IBT might look like. This represents scope for further research in the context of Beijing at least.

Lastly, our estimates of the deadweight loss show that there cash subsidies rather than subsidies through water consumption would be more efficient. This suggests that alternative schemes to assist the poor, such as the means tested rebate schemes underway in Chile, ought to be considered.

In sum there are a number of arguments which warn against using IBTs and which should be considered in more depth. Clearly, any efficiency loss associated with them needs to be weighed against the more equitable burden of tariffs that the IBT system has been demonstrated to induce in Beijing.

Appendices

1 Appendix 1: The Almost Ideal Demand System (AIDS)

The AIDS is founded upon the assumption that the indirect utility can be represented by the following general function of prices and income:

$$\ln V = \left[\frac{\ln y_h - \ln a(p)}{b(p)} \right] \tag{9}$$

where $a(p)$ and $b(p)$ are functions of the vector of prices and capture in a general way the nature of the response to price changes. In order to maintain flexibility in the estimated response these functions are usually represented in a flexible functional form. Deaton and Muellbauer (1980) characterise these responses as follows:

$$a(p) = \alpha_0 + \sum_{i=1}^{n} \alpha_i \ln p_i + \sum_{i=1}^{n} \sum_{j=1}^{n} \gamma_{ij} \ln p_i \ln p_j \tag{10}$$

$$b(p) = \prod_{i=1}^{n} p_i^{\beta_i} \tag{11}$$

where is commonly interpreted as the subsistence level of expenditure (Banks et al., 1997) and are the parameters to be estimated. It is easy to show that equations (9)–(11) yield the following Marshallian demands in budget shares form:

$$w_{ih} = a_i + \sum_{j} \gamma_{ij} \ln p_j + \beta_i \left[\ln y_h - a(p) \right] \tag{12}$$

where α_i is commonly set equal to the subsistence (log) income level (Banks et al., 1997). The AIDS and its successor the Quadratic Almost Ideal Demand System (QUAIDS) (Banks et al., 1997) are frequently used in consumption analysis in order to estimate consumer responses to price changes for nondurable goods such as food and clothes. The AIDS model gives an arbitrary second-order approximation to any demand system and provides a reasonably flexible representation of preferences. Deaton and Muellbauer (1980) show that it satisfies the axioms of choice, aggregates over consumers without a need to assume parallel Engel curves, and has a functional form consistent with known household budget data. In addition, it is simple to estimate and it can be used to test whether or not demand functions have the properties associated with utility maximisation such as homogeneity and symmetry.

The Rank of any demand system represents the maximum order of the approximation of the demand system, and hence the maximum dimension of the parameter space for the associated Engel curves (Lewbel, 1990). The AIDS is a Rank-2 demand system and therefore contains the restriction that the Engel curves are linear in the log of expenditure (Banks et al., 1997). Although Banks et al (1997) show that this restriction is indeed prohibitive for certain goods, our analysis of the Engel

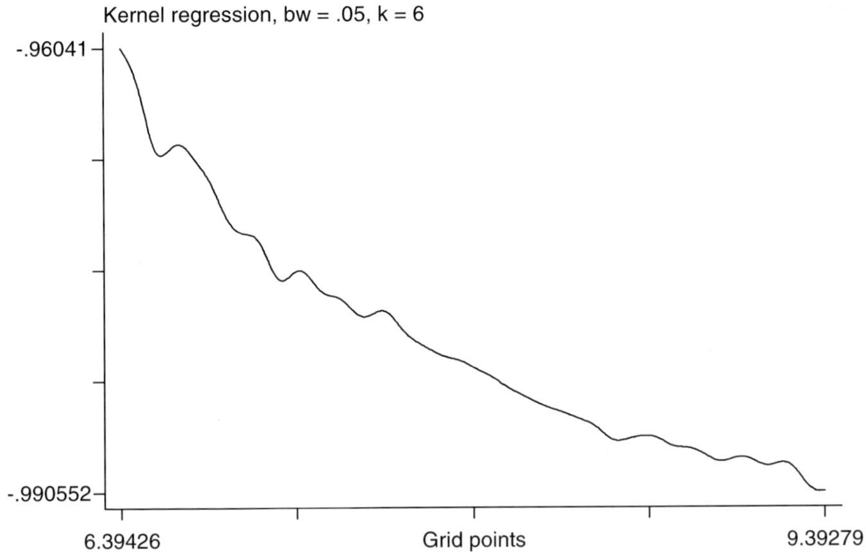

Fig. 9.A Semi-parametric Engel curve for water in Beijing

curve for water shown in Fig. 9.A below suggests that the AIDS is not restrictive in this context[25].

2 Appendix 2: Welfare Measures

The equality shown in (2) translates into the following equation in the AIDS system. And with p^* normalised to 1 the change in expenditure required to maintain welfare at the baseline levels after the price change can be written as the following expenditure index:

$$\frac{\ln x_h - \ln p}{b(p)} = \frac{\ln x_h^* - \ln p^*}{b(p^*)}$$

Taking $p^*=1$ yields the expenditure index:

$$\ln X_h = \ln\left(\frac{x_h}{x_h^*}\right) = \frac{a(p) + [b(p) - 1]\ln x_h}{b(p)} \tag{13}$$

[25] This Engel curve is estimated using a two-step semi-parametric approach similar to Robinson (1988). First we undertake a linear regression of budget share on prices. Then we undertake a kernel regression of the residuals of the OLS regression on income. In this way a non-parametric relation between budget share and the log of income is generated, controlled by a parametric relation between budget share and prices obtained in the first step.

The deadweight loss associated translates into the following measure in the AIDS model:

$$\ln W_h = \frac{\ln x_h^*(p) - \ln a(p)}{b(p)} - \frac{\ln x_h^* - \ln a(p^*)}{b(p^*)} \tag{14}$$

3 Appendix 3: Data Issues

Although the China HIES is one of the most comprehensive surveys in the world, there has been much discussion about the process of data collection and sample selection for both the rural and urban components of the Chinese HIES. Ravallion and Chen (1999) have made several observations concerning the quality of the data in the rural HIES. Their reservations about the quality of these data stem mainly from the manner in which the panel is constructed each year and the non-random attrition of households. Similarly, Gibson et al (2003) note with regard to the urban HIES that monthly reporting of expenditure is a somewhat onerous task for most households, a fact which means that recruiting respondents is particularly difficult. This, coupled with falling payments for participation in the survey, has potentially introduced some systematic bias to the sample selection since recruiting households to the survey is virtually impossible in certain areas. Non-random sample selection on the basis of personal, political or patriotic tendencies has frequently been adopted in response (Gibson et al., 2003).

The data on water expenditure and quantity are not uniformly spread through time since expenditures are recorded monthly and most households appear not to pay their water bill every month. Although normally the water authorities collect the water fees every month or every three months, the frequencies and the time of billing for different households may vary much. The reason could be that the residents are not at home when bill collectors check water meter and the residents could pay water bill over a long period at one time. This imbalance of frequency of billing for different households makes our estimation more complicated and to make households comparable expenditures are converted into monthly equivalents for the econometric analysis, whilst a variable reflecting the frequency of payments is constructed to reflect these differences.

4 Appendix 4: Non-linear pricing and predicting water demands

Fig. 9.B depicts a situation in which a price p_1 is charged for water demand between 0 and $\bar{w}$, and a price p_2 is charged for demand greater than $\bar{w}$. Where y is the Hicksian composite good, which, for simplicity, we assume, is the numeraire.

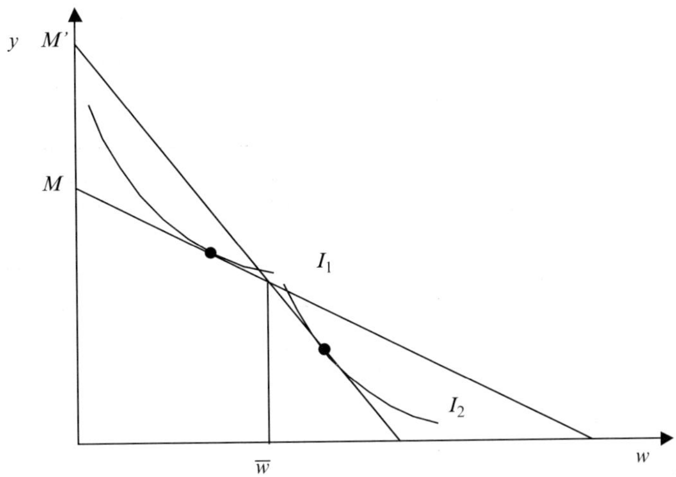

Fig. 9.B Non-linear budget sets under IBTs

Given this, the budget set faced by the household is kinked at $\overline{w}$. Therefore, consumption below $\overline{w}$.

$$M = p_1 w + y$$

For consumption above $\overline{w}$, the budget constraint can be described as follows:

$$M = (p_2 - p_1)\overline{w} + y + p_2 w$$

The latter case implies a 'virtual' budget constraint for consumption below $\overline{w}$ which is shown by the dotted line to M'. This 'virtual' budget can be written as:

$$M' = M + (p_2 - p_1)\overline{w} = p_2 w + y$$

It is clear that if the consumer demands more than $\overline{w}$ the budget constraint looks as if there has been an addition to income of the amount p_2y, complicating the maximisation problem. With this background the prediction procedure outlined in the text is easy to understand.

References

Banks, J., Blundell, R., and Weber. (1997). A quadratic almost ideal demand system. *Econometrica.*

Billings, R. B., and Agthe, D. E. (1980). Price elasticities for water: a case of increasing block rates. *Land Economics*, 56(1): 73–84.

Billings, R. (1982). Specification of block rate price variable in demand models. *Land Economics*, 58(3): 386–393.

Cavanagh, S.M., Hanemann, W.M., and Stavins, R.N. (2001). Muffled price signals: household water demand under increasing-block prices. FEEM Working Paper No. 40.2002.

CIDA Funded Project Team (1999). Report on urban household water use pilot survey in Beijing and Tianjin, Sustainable Water Resources Management in Beijing and Tianjin Region Water Demand Estimation. The centre for human settlements, University of British Columbia.

CCICED (2002). Implementing the Natural Forest Protection Programme and the Sloping Land Conversion Program: Lessons and Policy Recommendations. CCICED taskforce on forests and grasslands. China Forestry Publishing House.

Dalhuisen, J.M., Florax, R.J.G.M., Groot, H.L.F., and Nijkamp, P. (2001). Price and income elasticities of residential water demand. Tinbergen Institute Discussion Paper. TI 2001-057/3

Deaton, A. and Muellbauer. (1980). An almost ideal demand system. *Econometrica.*

Economy, E. (1997). Environmental Scarcities, State Capacity, and Civil Violence: The Case of China. (Part of the State Capacity project of the Toronto group, Homer-Dixon et al.)

Espey, M., Espey, J., and Shaw, W.D. (1997). Price elasticity of residential demand for water: a meta-analysis. *Water Resources Research* 33(6): 1369–1374.

Gibson, J., Huang, J., and Rozelle S. (2003). Improving estimates of Inequality and poverty from urban China's household income and expenditure survey. *Review of Income and Wealth*. 49(1): 53–68.

Herrington, P. (1995). Pricing water properly, en O'Riordan, T. Ecotaxation, 263: 286. Earthscan, London.

Höglund, L. (1999). Household demand for water in Sweden with implications of a potential tax on water use. *Water Resources Research*, 35(12): 3853–3863, 1999.

Hou, E. (2001). Nine Dragons, One River: The Role of Institutions in Developing Water Pricing Policy in Beijing, PRC. Full text documents. *3x3 Beijing Tianjin Water Resources Management Project*, The centre for human settlements, University of British Columbia.

Hou, E., and Hunter, G. (1998). Beijing Water: Causes, Effects, Solutions. Full text documents. *3x3 Beijing Tianjin Water Resources Management Project*. The centre for human settlements, University of British Columbia.

Hsiao, C. (2003). Analysis of Panel Data. Cambridge: Cambridge University Press.

Koundouri, P., Pashardes, P., and Hadjispirou, S. (2002). Equity and efficiency in water pricing: the case of Cyprus. *Environment and Development Economics.*

Lu Xiao-yan, Xu Fu-Liu, Shu Tao, Dawson, R. W., Gang Li Ben, Cao Jun, Leaf, M., and Hou Eve (2000). Current situation, defects and future suggestions of institutional framework for water resources management in China. Peking University Working paper.

Martínez-Espiñeira, R. (2002). Estimating Water Demand under Increasing Block tariffs Using Aggregate Data and Proportions of Users per Block. *Environmental and Resource Economics*, 26: 5–23, 2003.

Moffitt, R. (1986). The econometrics of piecewise-linear budget constraints: A survey and exposition of the Maximum Likelihood Method. *Journal of Business & Economics Statistics*, July 1986, 4(3): 317–328.

Nieswiadomy, M.L. and Molina, D.J. (1991). A note on price perception in water demand models. *Land Economics* 67(3): 352–359.

Nordin, J. (1976). A proposed modification on Taylor's demand supply analysis: comment. Bell *Journal of Economic Management and Science*, 7(2): 719–721.

OECD, (1999b). Household water pricing in OECD countries. OECD, Paris, unclassified document ENV/EPOC/GEEI(98) 12/FINAL.

Planning Report (2001). Sustainable Utilization Planning of Capital Water Resource in the Early Stage of the 21st Century (2001–2005). Government of China.

Seery, M. (1996). Distributional impact of water pricing. Equity and the Environment.

Smith, S. and Rajah, N. (1993). Distributional aspects of household water charges. *Fiscal Studies*, 14(3): 86–108.

Sterner, T. (2003). Policy Instruments for Environmental and Natural Resource Management. Resources for the Future. Washington DC.
US Water News (2004). Water Hikes Expected in Beijing. US water News online, June. Article can be found at:
www.uswaternews.com/archives/arcglobal/4waterpric6.html
Whittington, D. and Boland, B. (2000). The political economy of Increasing block tariffs. Ch 10 in Dinar (2000) the political economy of water pricing.
Whittington, D. (2002). Municipal Water Pricing and Tariff Design: A Reform Agenda for cities in Developing countries. Resources for the Future.
World Bank (1997). Water pricing experiences: An international perspective.
Xinhuanet (2004). www.xinhuanet.com, 2004–06–16.
Yan P., and Stover, J. (1999). Water pricing issues and potential legal reforms in China. *Sinosphere* 2(4).
Yatchew. (1998). Non and semi-parametric estimation techniques: A review. *Journal of Economic Surveys*.
Yepes, G. (1999). Do Cross-subsidies help the poor to benefit from water and wastewater services Lessons from Guayaquil. World Bank.

Index

ENVIRONMENT & POLICY

1. Dutch Committee for Long-Term Environmental Policy: *The Environment: Towards a Sustainable Future*. 1994
ISBN 0-7923-2655-5; Pb 0-7923-2656-3
2. O. Kuik, P. Peters and N. Schrijver (eds.): *Joint Implementation to Curb Climate Change*. Legal and Economic Aspects. 1994 ISBN 0-7923-2825-6
3 C.J. Jepma (ed.): *The Feasibility of Joint Implementation*. 1995
ISBN 0-7923-3426-4
4 F.J. Dietz, H.R.J. Vollebergh and J.L. de Vries (eds.): *Environment, Incentives and the Common Market*. 1995 ISBN 0-7923-3602-X
5. J.F.Th. Schoute, P.A. Finke, F.R.Veeneklaas and H.P.Wolfert (eds.): *Scenario Studies for the Rural Environment*. 1995 ISBN 0-7923-3748-4
6. R.E. Munn, J.W.M. la Rivière and N. van Lookeren Campagne: *Policy Making in an Era of Global Environmental Change*. 1996
ISBN 0-7923-3872-3
7. F. Oosterhuis, F. Rubik and G. Scholl: *Product Policy in Europe: New Environmental Perspectives*. 1996 ISBN 0-7923-4078-7
8. J. Gupta: *The Climate Change Convention and Developing Countries: From Conflict to Consensus?* 1997 ISBN 0-7923-4577-0
9. M. Rolén, H. Sjöberg and U. Svedin (eds.): *International Governance on Environmental Issues*. 1997 ISBN 0-7923-4701-3
10. M.A. Ridley: *Lowering the Cost of Emission Reduction: Joint Implementation in the Framework Convention on Climate Change*. 1998
ISBN 0-7923-4914-8
11. G.J.I. Schrama (ed.): *Drinking Water Supply and Agricultural Pollution*. Preventive Action by the Water Supply Sector in the European Union and the United States. 1998 ISBN 0-7923-5104-5
12. P. Glasbergen: *Co-operative Environmental Governance:* Public-Private Agreements as a Policy Strategy. 1998.
ISBN 0-7923-5148-7; Pb 0-7923-5149-5
13. P. Vellinga, F. Berkhout and J. Gupta (eds.): *Managing a Material World*. Perspectives in Industrial Ecology. 1998
ISBN 0-7923-5153-3; Pb 0-7923-5206-8
14. F.H.J.M. Coenen, D. Huitema and L.J. O'Toole, Jr. (eds.): *Participation and the Quality of Environmental Decision Making*. 1998 ISBN 0-7923-5264-5
15. D.M. Pugh and J.V. Tarazona (eds.): *Regulation for Chemical Safety in Europe: Analysis, Comment and Criticism*. 1998 ISBN 0-7923-5269-6
16. W. Østreng (ed.): *National Security and International Environmental Cooperation in the Arctic – the Case of the Northern Sea Route*. 1999
ISBN 0-7923-5528-8
17. S.V. Meijerink: *Conflict and Cooperation on the Scheldt River Basin*. A Case Study of Decision Making on International Scheldt Issues between 1967 and 1997. 1999 ISBN 0-7923-5650-0

ENVIRONMENT & POLICY

18. M.A. Mohamed Salih: *Environmental Politics and Liberation in Contemporary Africa.* 1999 ISBN 0-7923-5652-7
19. C.J. Jepma and W. van der Gaast (eds.): *On the Compatibility of Flexible Instruments.* 1999 ISBN 0-7923-5728-0
20. M. Andersson: *Change and Continuity in Poland's Environmental Policy.* 1999 ISBN 0-7923-6051-6
21. W. Kägi: *Economics of Climate Change: The Contribution of Forestry Projects.* 2000 ISBN 0-7923-6103-2
22. E. van der Voet, J.B. Guinée and H.A.U. de Haes (eds.): *Heavy Metals: A Problem Solved?* Methods and Models to Evaluate Policy Strategies for Heavy Metals. 2000 ISBN 0-7923-6192-X
23. G. Hønneland: *Coercive and Discursive Compliance Mechanisms in the Management of Natural Resourses.* A Case Study from the Barents Sea Fisheries. 2000 ISBN 0-7923-6243-8
24. J. van Tatenhove, B. Arts and P. Leroy (eds.): *Political Modernisation and the Environment.* The Renewal of Environmental Policy Arangements. 2000 ISBN 0-7923-6312-4
25. G.K. Rosendal: *The Convention on Biological Diversity and Developing Countries.* 2000 ISBN 0-7923-6375-2
26. G.H. Vonkeman (ed.): *Sustainable Development of European Cities and Regions.* 2000 ISBN 0-7923-6423-6
27. J. Gupta and M. Grubb (eds.): *Climate Change and European Leadership.* A Sustainable Role for Europe? 2000 ISBN 0-7923-6466-X
28. D. Vidas (ed.): *Implementing the Environmental Protection Regime for the Antarctic.* 2000 ISBN 0-7923-6609-3; Pb 0-7923-6610-7
29. K. Eder and M. Kousis (eds.): *Environmental Politics in Southern Europe:* Actors, Institutions and Discourses in a Europeanzing Society. 2000 ISBN 0-7923-6753-7
30. R. Schwarze: *Law and Economics of International Climate Change Policy.* 2001 ISBN 0-7923-6800-2
31. M.J. Scoullos, G.H. Vonkeman, I. Thornton, and Z. Makuch: *Mercury - Cadmium-Lead:* Handbook for Sustainable Heavy Metals Policy and Regulation. 2001 ISBN 1-4020-0224-6
32. G. Sundqvist: *The Bedrock of Opinion. Science,* Technology and Society in the Siting of High-Level Nuclear Waste. 2002 ISBN 1-4020-0477-X
33. P.P.J. Driessen and P. Glasbergen (eds.): *Greening Society.* The Paradigm Shift in Dutch Environmental Politics. 2002 ISBN 1-4020-0652-7
34. D. Huitema: *Hazardous Decisions.* Hazardous Waste Siting in the UK, The Netherlands and Canada. Institutions and Discourses. 2002 ISBN 1-4020-0969-0
35. D.A. Fuchs: *An Institutional Basis for Environmental Stewardship:* The Structure and Quality of Property Rights. 2003 ISBN 1-4020-1002-8

ENVIRONMENT & POLICY

36. B. Chaytor and K.R. Gray (eds.): *International Environmental Law and Policy in Africa.* 2003 ISBN 1-4020-1287-X
37. F.M. Brouwer, I. Heinz and T. Zabel (eds.): *Governance of Water-Related Conflicts in Agriculture.* New Directions in Agri-Environmental and Water Policies in the EU. 2003 ISBN 1-4020-1553-4
38. G.J.I. Schrama and S. Sedlacek (eds.): *Environmental and Technology Policy in Europe.* Technological Innovation and Policy Integration. 2003 ISBN 1-4020-1583-6
39. A.J. Dietz, R. Ruben and A. Verhagen (eds.): *The Impact of Climate Change on Drylands.* With a focus on West Africa. 2004 ISBN 1-4020-1952-1
40. I. Kissling-Näf and S. Kuks (eds.): *The Evolution of National Water Regimes in Europe.* Transitions in Water Rights and Water Policies. 2004 ISBN 1-4020-2483-5
41. H. Bressers and S.Kuks (eds.): *Integrated Governance and Water Basin Management.* Conditions for Regime Change and Sustainability. 2004 ISBN 1-4020-2481-9
42. T. Flüeler: *Decision Making for Complex Socio-Technical Systems.* Robustness from Lessons Learned in Long-term Radioactive Waste Governance. 2005 ISBN 1-4020-3480-6
43. E. Croci (ed.): *The Handbook of Environmental Voluntary Agreements.* Design, Implementation and Evaluation Issues. 2005 ISBN 1-4020-3355-9
44. X. Olsthoorn and A.J. Wieczorek (eds.): *Understanding Industrial Transformation.* Views from Different Disciplines. 2006 ISBN 1-4020-3755-4
45. H. Aiking, J. de Boer and J. Vereijken (eds.): *Sustainable Protein Production and Consumption: Pigs or Peas?* 2006 ISBN 1-4020-4062-8
46. F. Brouwer and B.A. McCarl (eds.): *Agriculture and Climate Beyond 2015.* A New Perspective on Future Land Use Patterns. 2006 ISBN 1-4020-4063-6
47. B. Arts and P. Leroy (eds.): *Institutional Dynamics in Environmental Governance.* 2006 ISBN 1-4020-5078-X
48. P. Koundouri (ed.): *Coping with Water Deficiency.* From Research to Policymaking. 2008 ISBN 978-1-4020-6614-6